Sum Fun
Maths Assessment

Years 3–4
Maths Assessment Puzzles for the 2014 Curriculum

Katherine Bennett

We hope you and your pupils enjoy solving the maths puzzles in this book. Brilliant Publications publishes other books for maths and maths problems. To find out more details on any of the titles listed below, please log onto our website: www.brilliantpublications.co.uk.

Maths Problem Solving Year 1	978-1-903853-74-0
Maths Problem Solving Year 2	978-1-903853-75-7
Maths Problem Solving Year 3	978-1-903853-76-4
Maths Problem Solving Year 4	978-1-903853-77-1
Maths Problem Solving Year 5	978-1-903853-78-8
Maths Problem Solving Year 6	978-1-903853-79-5

Maths Problems and Investigations 5–7 year olds	978-0-85747-626-5
Maths Problems and Investigations 7–9 year olds	978-0-85747-627-2
Maths Problems and Investigations 9–11 year olds	978-0-85747-628-9

The Mighty Multiples Times Table Challenge 978-0-85747-629-6

Published by Brilliant Publications
Unit 10
Sparrow Hall Farm
Edlesborough
Dunstable
Bedfordshire
LU6 2ES, UK

E-mail:
info@brilliantpublications.co.uk
Website:
www.brilliantpublications.co.uk
Tel: 01525 222292

Written by Katherine Bennett
Illustrated by Gaynor Berry
Front cover illustration by Gaynor Berry

Printed ISBN 978-78317-084-5
e-book ISBN 978-78317-089-0

First printed and published in the UK in 2014

Contents

Introduction

The aim of the 'Sum Fun' series is to enable teachers to gather evidence and assess children's learning in maths.

Linked to year group objectives from the new September 2014 curriculum, each fun activity sheet requires pupils to use their mathematical skills to solve a series of questions. They must then use the answers to 'crack the code' and find the solutions to silly jokes, puns and riddles. The activities use Assessment for Learning techniques, such as child friendly 'I can …' statements at the top of each sheet, so that pupils can be clear about the learning objective; they also encourage self-assessment because if a solution doesn't make sense, pupils will need to spot and correct their mistakes. Quick reference answer pages are provided for the teacher at the back of the book, or to enable pupils to self-mark. There are several sheets per objective so that each one can be tested at different points in the year if necessary, without repetition of the same questions and jokes. This could be at the end of a unit of work, or as a one-off assessment task. The assessment checklist on page 106 will help you to keep track of children's progress.

The activities are in a fun format that children soon become familiar with and look forward to solving, promoting high levels of pupil engagement. Children are motivated by the fun element of the jokes and will compete to be the first to get the answer!

As well as an assessment tool, the sheets can be used as independent tasks in everyday lessons. They are clearly linked to year group objectives from the new curriculum, providing an easy way of differentiating group or individual activities without any extra work for the class teacher! They make good whole class starter or plenary activities on an interactive whiteboard, or could just be used as fun 'time fillers'!

Sequences: 4, 8, 50 and 100 (1)

Learning objectives
I can count forwards and backwards in multiples of 4, 8, 50 and 100.

+ 3 ×
¾ = 8
1.5 ÷

To solve the jokes, work out the number that comes next and write it in the oval. Then use the grid to find the letter that goes with each answer and write it on the line. The first one is done for you!

64	88	250	200	24	400	104	16	500	60	72	750
S	K	W	R	H	A	E	D	N	L	I	T

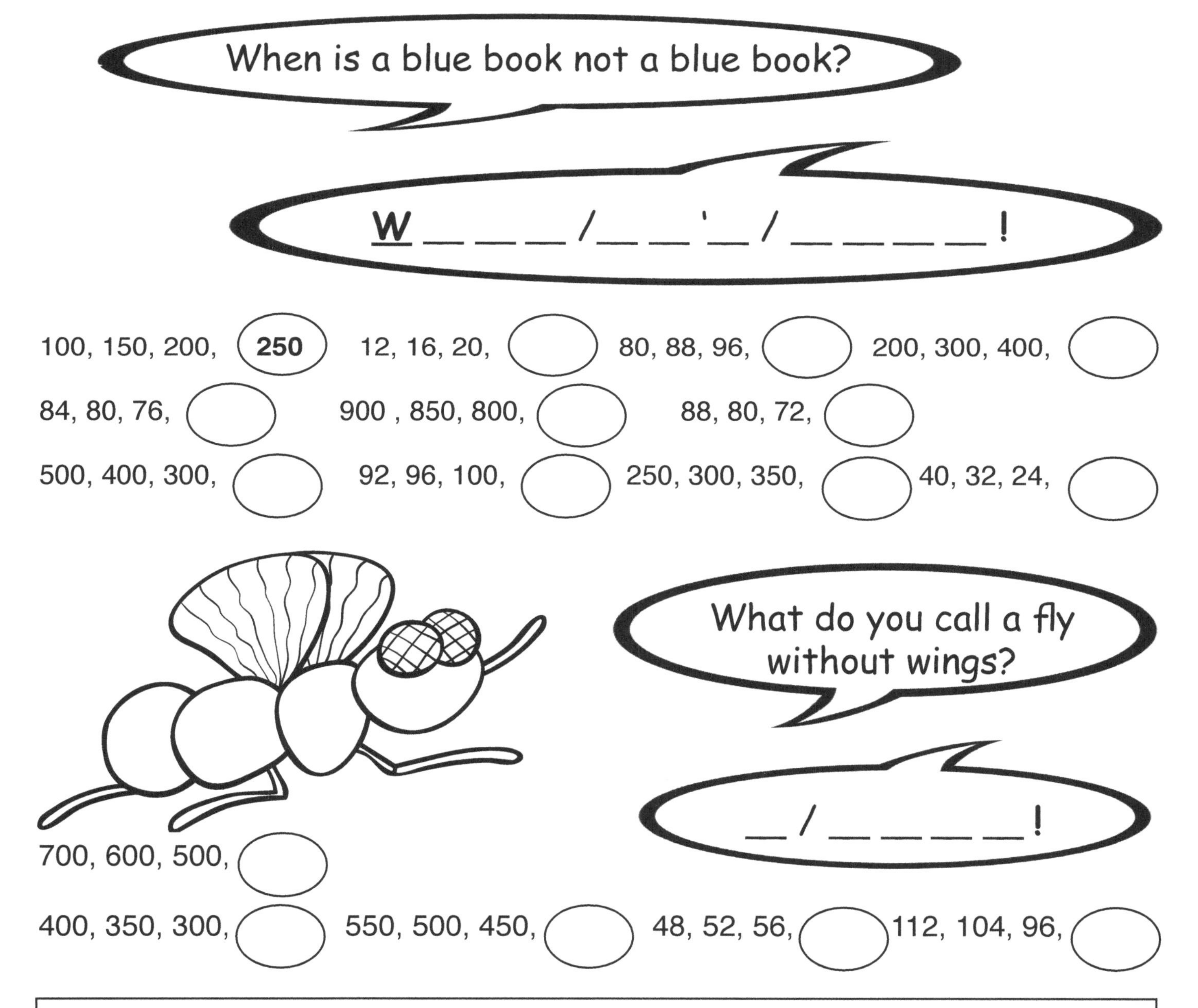

Year 3 – Number and place value
- *Count from 0 in multiples of 4, 8, 50 and 100.*

Sequences: 4, 8, 50 and 100 (2)

Learning objectives
I can count forwards and backwards in multiples of 4, 8, 50 and 100.

+ 3 ×
¾ = 8
1.5 ÷

To solve the jokes, work out the number that comes next and write it in the oval. Then use the grid to find the letter that goes with each answer and write it on the line. The first one is done for you!

300	48	750	600	64	850	96	24	150	32	550	0
E	W	A	M	L	Y	O	D	F	R	T	H

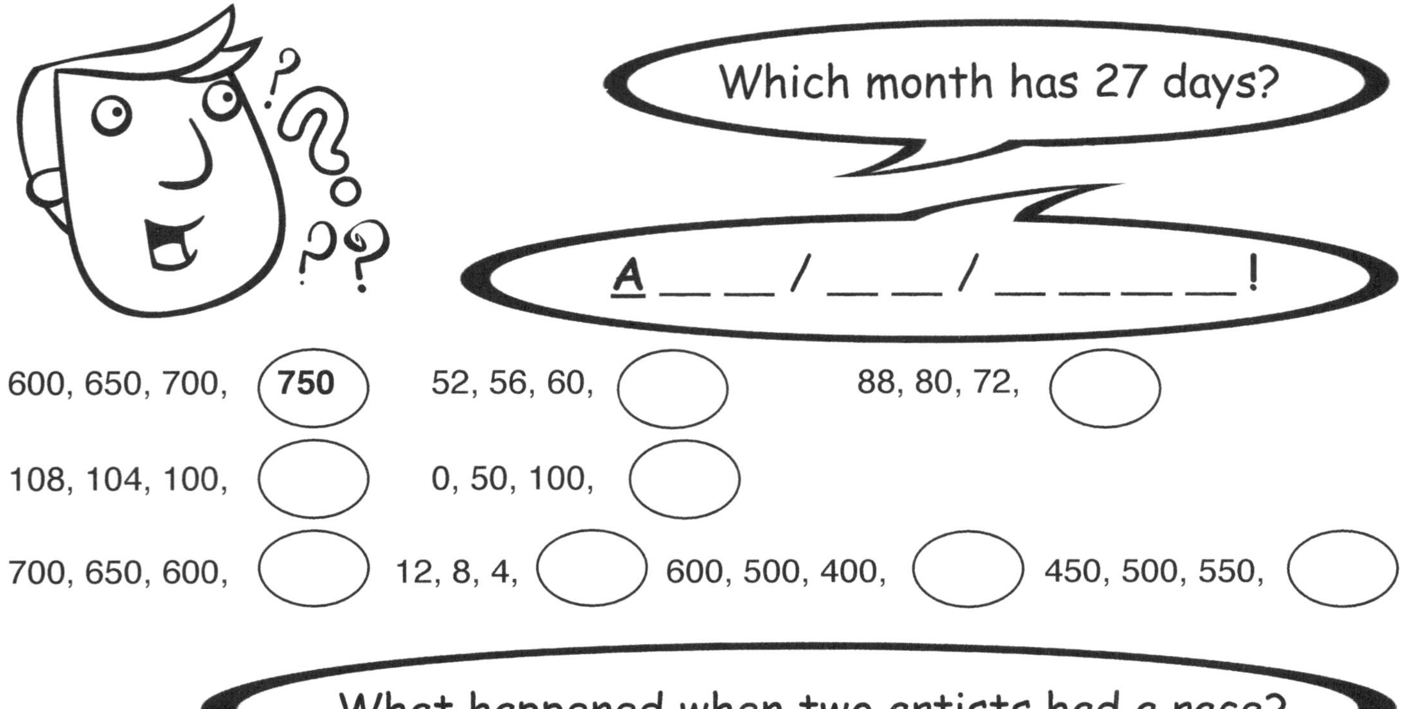

600, 650, 700, (**750**) 52, 56, 60, ( ) 88, 80, 72, ( )

108, 104, 100, ( ) 0, 50, 100, ( )

700, 650, 600, ( ) 12, 8, 4, ( ) 600, 500, 400, ( ) 450, 500, 550, ( )

What happened when two artists had a race?

_ _ _ _ _ / _ _ _ _ _ !

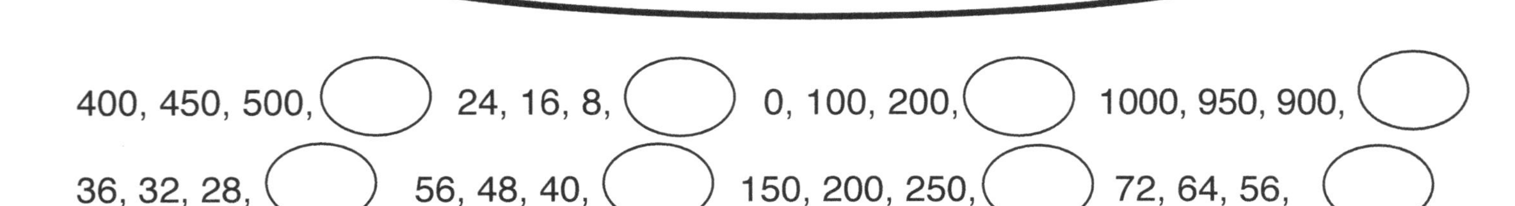

400, 450, 500, ( ) 24, 16, 8, ( ) 0, 100, 200, ( ) 1000, 950, 900, ( )

36, 32, 28, ( ) 56, 48, 40, ( ) 150, 200, 250, ( ) 72, 64, 56, ( )

Year 3 – Number and place value
- *Count from 0 in multiples of 4, 8, 50 and 100.*

Add and subtract 10 or 100 (1)

Learning objectives
I can count 10 or 100 more than a number.
I can count 10 or 100 less than a number.

To solve the jokes, work out the answer to the question and write it in the oval. Then use the grid to find the letter that goes with each answer and write it on the line. The first one is done for you!

595	258	395	819	458	82	609	782
I	B	D	O	L	N	Y	E

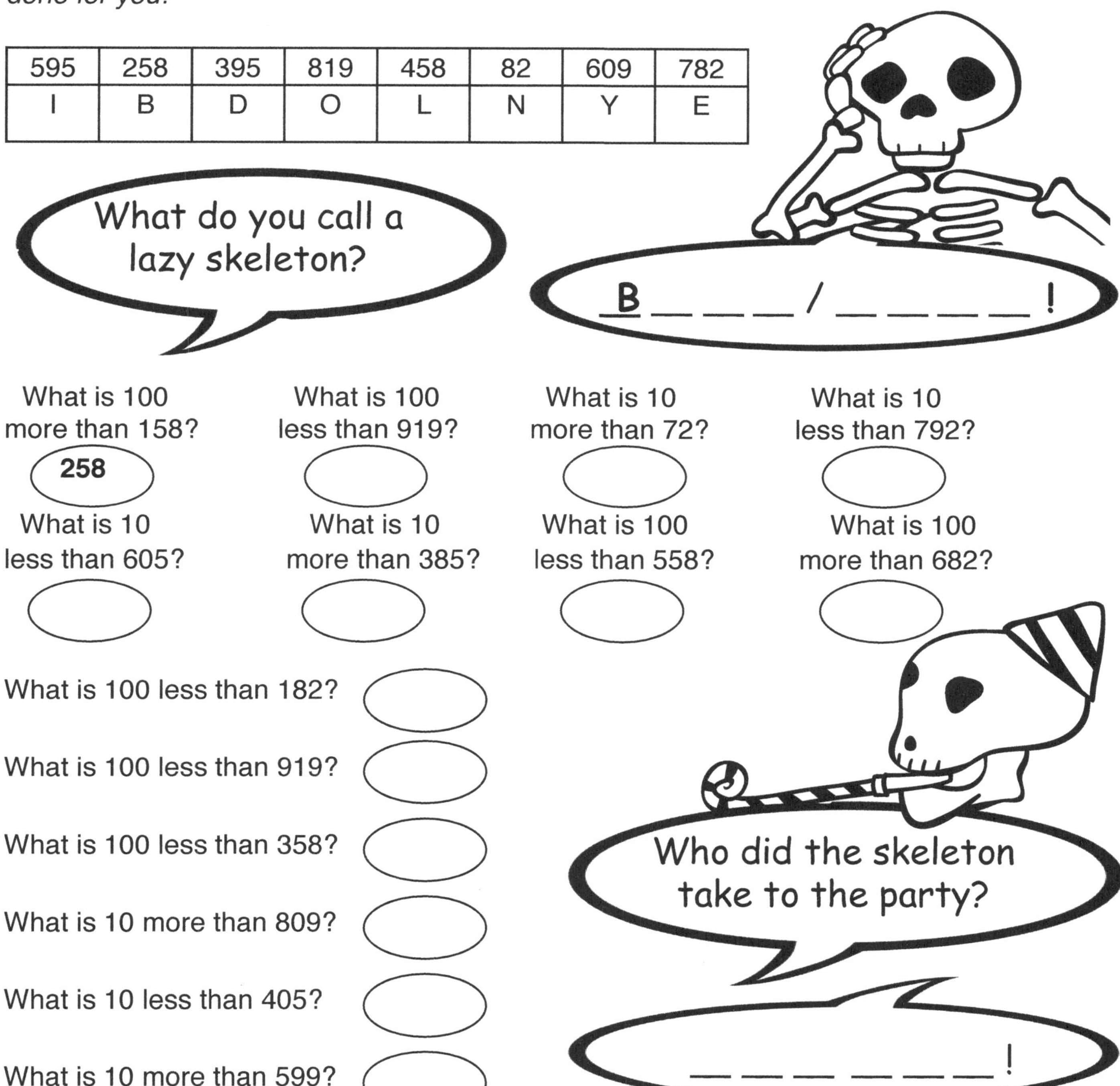

What is 100 more than 158? **258**

What is 100 less than 919?

What is 10 more than 72?

What is 10 less than 792?

What is 10 less than 605?

What is 10 more than 385?

What is 100 less than 558?

What is 100 more than 682?

What is 100 less than 182?

What is 100 less than 919?

What is 100 less than 358?

What is 10 more than 809?

What is 10 less than 405?

What is 10 more than 599?

Year3 – Number and place value
- *Find 10 or 100 more or less than a given number.*

Add and subtract 10 or 100 (2)

Learning objectives
I can count 10 or 100 more than a number.
I can count 10 or 100 less than a number.

To solve the joke, work out the answer to the question and write it in the oval. Then use the grid to find the letter that goes with each answer and write it on the line. The first one is done for you!

208	491	391	534	416	218	252	734	607	52	405	691	674
M	B	I	Y	L	S	H	C	A	U	R	T	E

How do rabbits send letters?

B __ / __ __ __ __ __ /
__ __ __ __ __ !

What is 100 more than 391?

491

What is 100 less than 634?

What is 100 more than 152?

What is 10 more than 597?

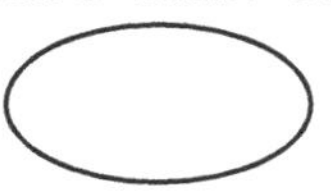

What is 10 less than 415?

What is 100 more than 574?

What is 10 more than 198?

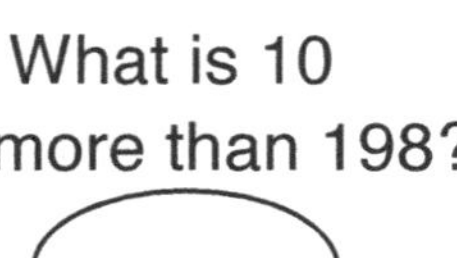

What is 10 less than 617?

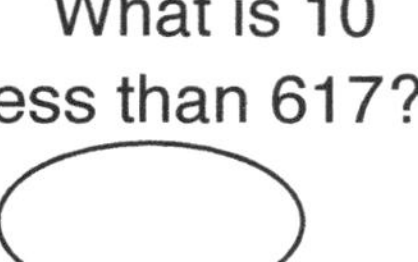

What is 10 less than 401?

What is 10 more than 406?

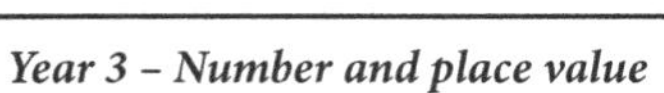
Year 3 – Number and place value
- *Find 10 or 100 more or less than a given number.*

Add and subtract 10 or 100 (3)

Learning objectives
I can count 10 or 100 more than a number.
I can count 10 or 100 less than a number.

To solve the joke, work out the answer to the question and write it in the oval. Then use the grid to find the letter that goes with each answer and write it on the line. The first one is done for you!

208	491	391	534	416	218	252	734	607	52	405	691	674
M	B	I	Y	L	S	H	C	A	U	R	T	E

What is 10 less than 262? **252**

What is 100 less than 707?

What is 10 more than 395?

What is 10 more than 664?

What is 10 more than 724?

What is 100 less than 152?

What is 10 less than 701?

What is 10 more than 208?

Year 3 – Number and place value
- *Find 10 or 100 more or less than a given number.*

Place value hundreds, tens and ones (1)

Learning objectives
I know the value of each digit in numbers up to 1000.

To solve the jokes, use place value to work out the value of the underlined digit and write the answer in the oval. Then use the grid to find the letter that goes with each answer and write it on the line. The first one is done for you!

30	8	50	1	40	6	300	70	9	4	100	5	3
L	H	N	A	M	F	U	P	I	D	S	E	C

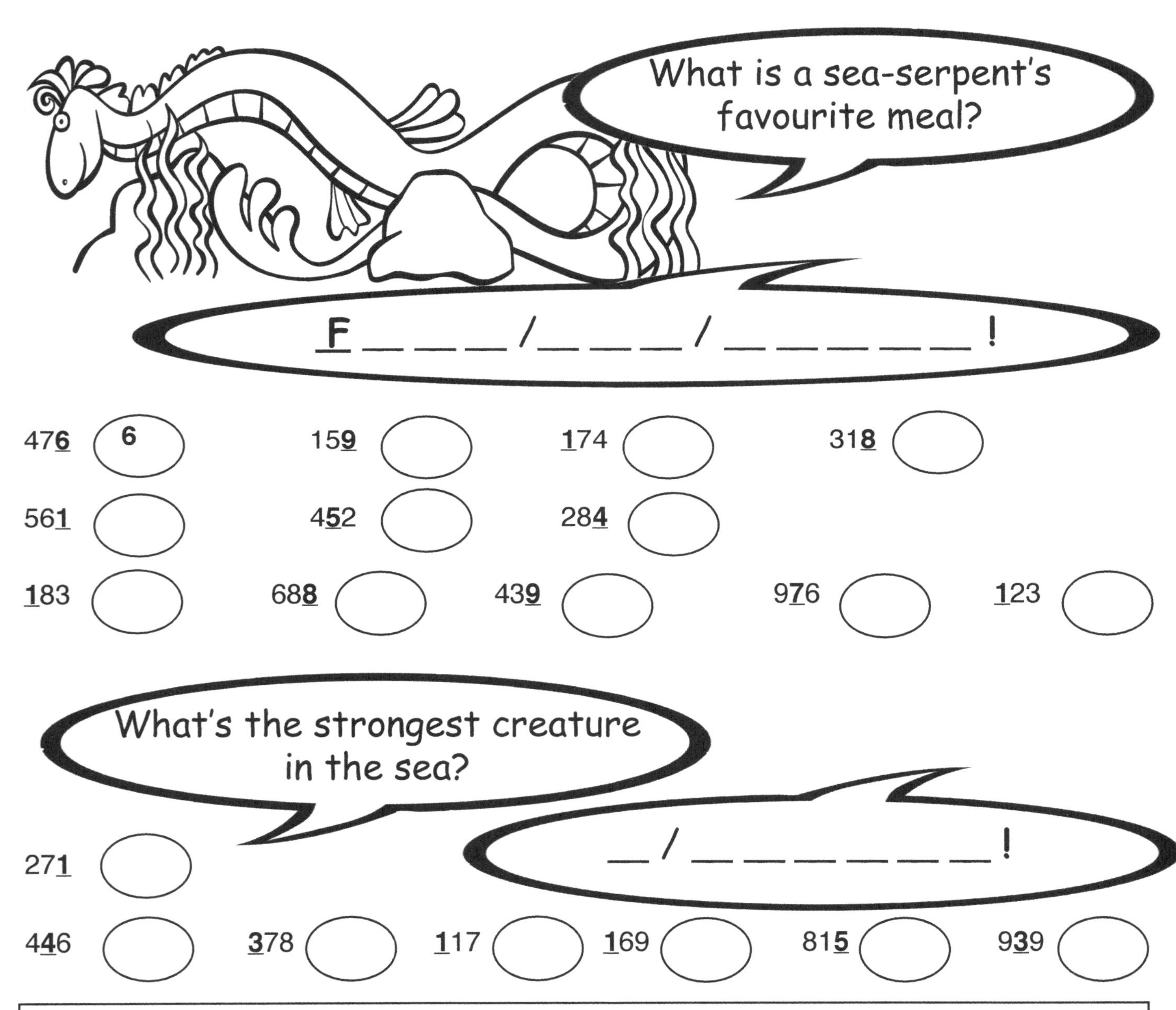

Year 3 – Number and place value

- *Recognise the place value of each digit in a three-digit number (hundreds, tens, ones).*

Place value hundreds, tens and ones (2)

Learning objectives
I know the value of each digit in numbers up to 1000.

+ 3 ×
¾ = 8
1.5 ÷

To solve the jokes, use place value to work out the value of the underlined digit and write the answer in the oval. Then use the grid to find the letter that goes with each answer and write it on the line. The first one is done for you!

70	100	50	90	800	9	40	2	30	5	3	1	300
L	S	O	R	T	Y	M	B	F	E	C	A	U

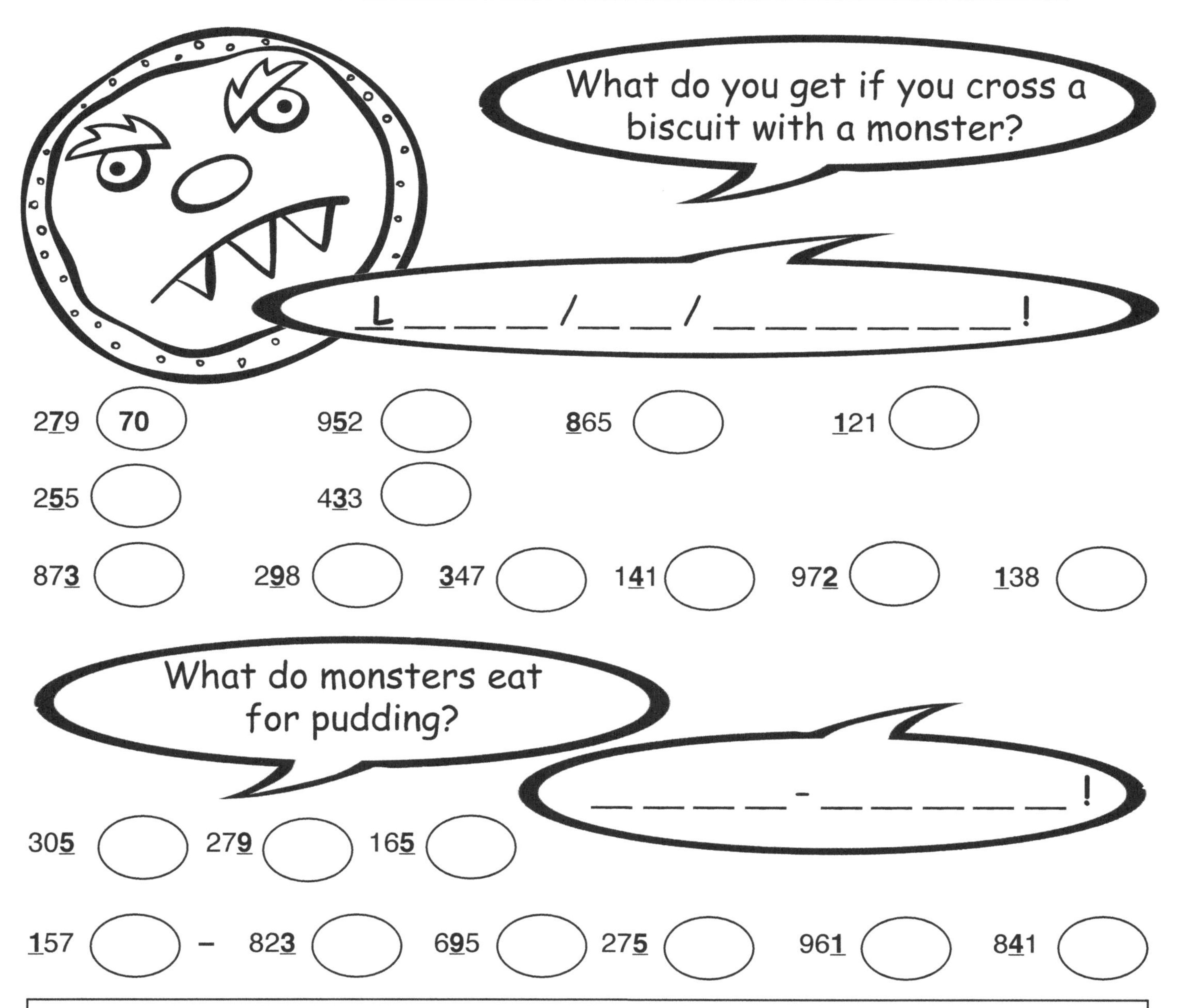

2**7**9 (70) 9**5**2 ( ) **8**65 ( ) **1**21 ( )

2**5**5 ( ) 4**3**3 ( )

87**3** ( ) 2**9**8 ( ) **3**47 ( ) 1**4**1 ( ) 97**2** ( ) **1**38 ( )

30**5** ( ) 27**9** ( ) 16**5** ( )

157 ( ) – 82**3** ( ) 6**9**5 ( ) 27**5** ( ) 96**1** ( ) 8**4**1 ( )

Year 3 – Number and place value
- *Recognise the place value of each digit in a three-digit number (hundreds, tens, ones).*

Reading numbers (1)

Learning objectives
I can read and understand a number from 1 to 1000 written in words.
I can write numbers from 1 to 1000 in words.

+ 3 ×
¾ = 8
1.5 ÷

To solve the joke, read the words and write each one as a number in the oval, then use the grid to find the letter that goes with each answer and write it on the line. The first one is done for you!

313	560	1000	330	281	420	218	516	468	280
H	A	Q	N	U	E	C	R	K	T

What would happen if all the ducks in the world jumped up and down at the same time?

A __ / __ __ __ __ __ __ __ / __ __ __ __ __ __ __ !

five hundred and sixty (560)

three hundred and thirty

four hundred and twenty

five hundred and sixty

five hundred and sixteen

two hundred and eighty

three hundred and thirteen

one thousand

two hundred and eighty-one

five hundred and sixty

two hundred and eighteen

four hundred and sixty-eight

Year 3 – Number and place value
- *Read and write numbers up to 1000 in numerals and in words.*

Reading numbers (2)

Learning objectives
I can read and understand a number from 1 to 1000 written in words.
I can write numbers from 1 to 1000 in words.

To solve the joke, read the words and write each one as a number in the ovals, then use the grid to find the letter that goes with each answer and write it on the line. The first one is done for you!

670	958	985	617	240	214	241
B	E	L	S	O	D	T

six hundred and seventy

nine hundred and eighty five

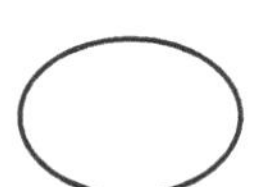

two hundred and forty

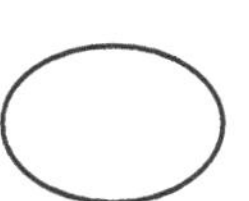

two hundred and forty

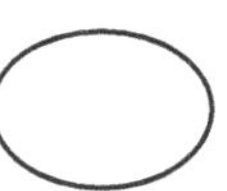

two hundred and fourteen

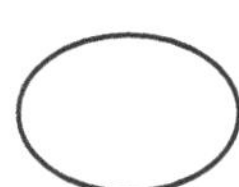

two hundred and forty-one

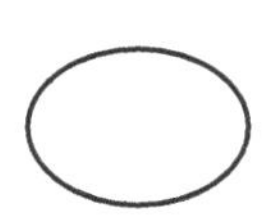

nine hundred and fifty-eight

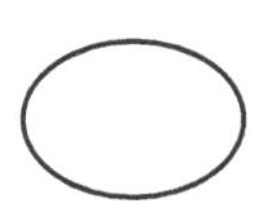

six hundred and seventeen

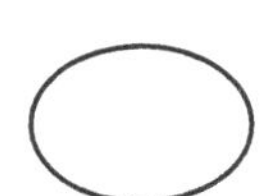

two hundred and forty-one

six hundred and seventeen

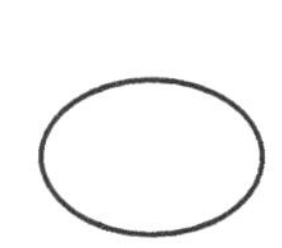

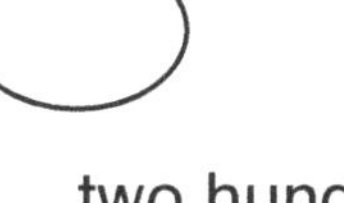

Year 3 – Number and place value
- *Read and write numbers up to 1000 in numerals and in words.*

Writing numbers (1)

Learning objectives
I can read and understand a number from 1 to 1000 written in words.
I can write numbers from 1 to 1000 in words.

This time, write the numbers as words on the line. Then match the answer to the grid to solve the joke. The first one is done for you!

seven hundred and eight	three hundred and twenty-two	seven hundred and eighty	three hundred and twelve	six hundred and eighteen	three hundred and twenty	seven hundred and eighteen	three hundred and two
Q	E	U	G	A	S	C	K

What's the first thing a duck does when it makes an omelette?

Q _ _ _ _ _ / _ _ _ _ _ _ !

708 **seven hundred and eight**

780 ______

618 ______

718 ______

302 ______

322 ______

312 ______

312 ______

320 ______

Year 3 – Number and place value
- *Read and write numbers up to 1000 in numerals and in words.*

Writing numbers (2)

Learning objectives
I can read and understand a number from 1 to 1000 written in words.
I can write numbers from 1 to 1000 in words.

This time, write the numbers as words on the line. Then match the answer to the grid to solve the joke. The first one is done for you!

four hundred and thirty	five hundred and fifteen	four hundred and thirteen	five hundred and fifty	four hundred and thirty-one	five hundred and fifty-one
A	H	C	O	T	E

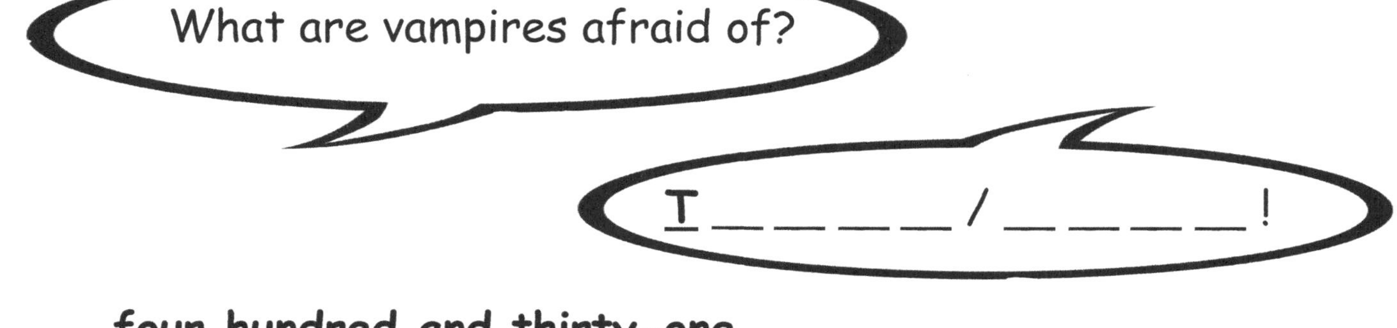

431 **four hundred and thirty-one** ________________

550 ________________

550 ________________

431 ________________

515 ________________

430 ________________

413 ________________

515 ________________

551 ________________

Year 3 – Number and place value
- *Read and write numbers up to 1000 in numerals and in words.*

Add and subtract ones, tens and hundreds (1)

Learning objectives
I can add ones, tens and hundreds to a 3-digit number.
I can subtract ones, tens and hundreds from a 3-digit number.

To solve the joke, write the answer to the maths question in the oval. Then use the grid to find the letter that goes with each answer and write it on the line. The first one is done for you!

297	158	753	482	567	138	677	446	517	933	59	828
H	Y	I	B	A	O	F	N	S	E	U	C

442 + 40 = **482**　　338 – 200 = ◯　　449 – 3 = ◯

133 + 800 = ◯

228 + 600 = ◯　　305 – 8 = ◯　　693 + 60 = ◯

416 + 30 = ◯　　867 – 300 = ◯

Year 3 – Addition and subtraction
- *Add and subtract numbers mentally, including:*
  - *a three-digit number and ones*
  - *a three-digit number and tens*
  - *a three-digit number and hundreds.*

Add and subtract ones, tens and hundreds (2)

Learning objectives
I can add ones, tens and hundreds to a 3-digit number.
I can subtract ones, tens and hundreds from a 3-digit number.

To solve the joke, write the answer to the maths question in the oval. Then use the grid to find the letter that goes with each answer and write it on the line. The first one is done for you!

297	158	753	482	567	138	677	446	517	933	59	828
H	Y	I	B	A	O	F	N	S	E	U	C

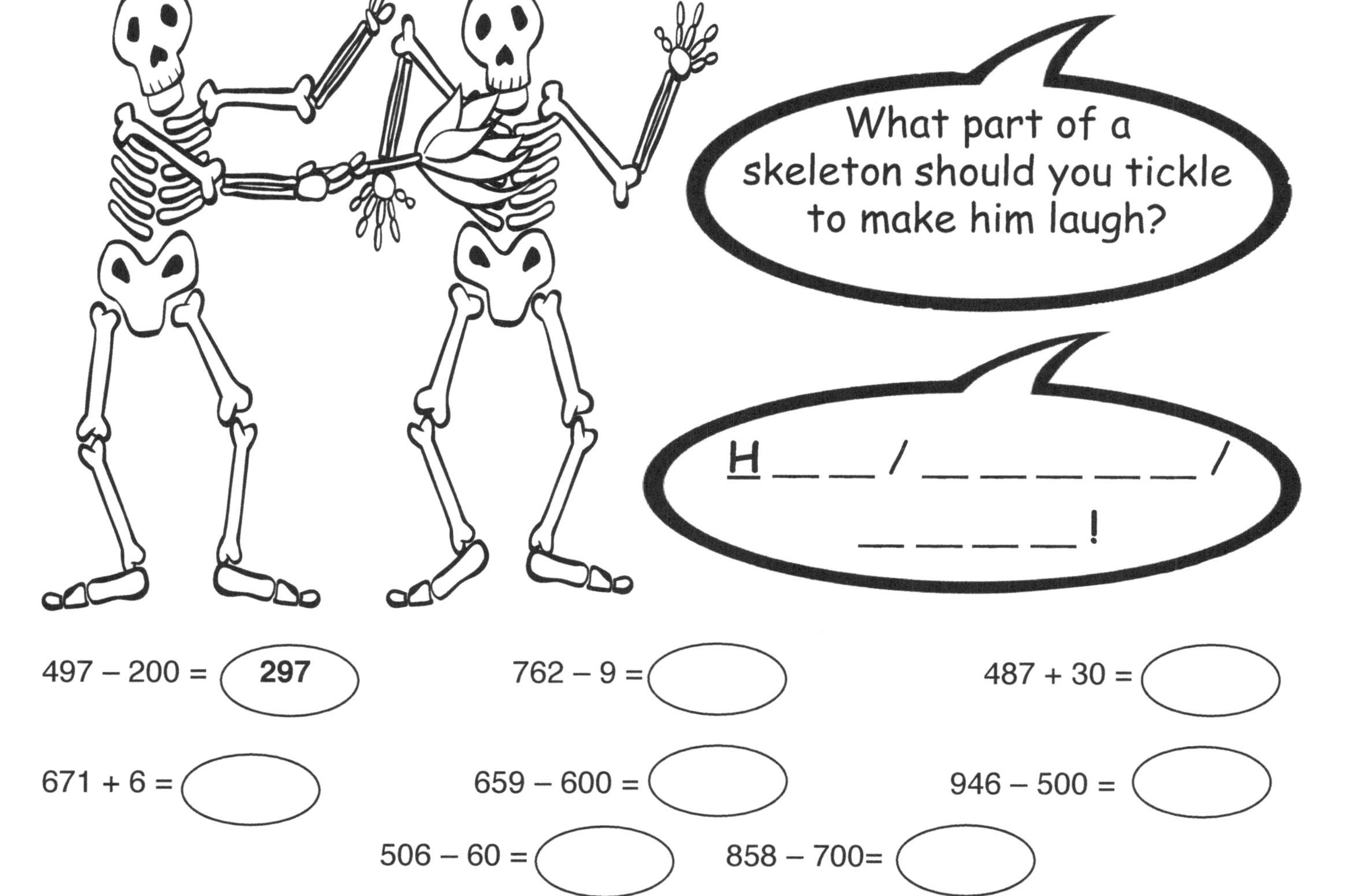

497 – 200 = **297**　　762 – 9 = ◯　　487 + 30 = ◯

671 + 6 = ◯　　659 – 600 = ◯　　946 – 500 = ◯

506 – 60 = ◯　　858 – 700= ◯

442 + 40 = ◯　　188 – 50 = ◯　　439 + 7 = ◯　　893 + 40 = ◯

Year 3 – Addition and subtraction
- *Add and subtract numbers mentally, including:*
  - *a three-digit number and ones*
  - *a three-digit number and tens*
  - *a three-digit number and hundreds.*

Add and subtract ones, tens and hundreds (3)

Learning objectives
I can add ones, tens and hundreds to a 3-digit number.
I can subtract ones, tens and hundreds from a 3-digit number.

To solve the joke, write the answer to the maths question in the oval. Then use the grid to find the letter that goes with each answer and write it on the line. The first one is done for you!

713	452	537	198	627	454	562	929	67	98
A	R	W	O	T	U	C	G	D	H

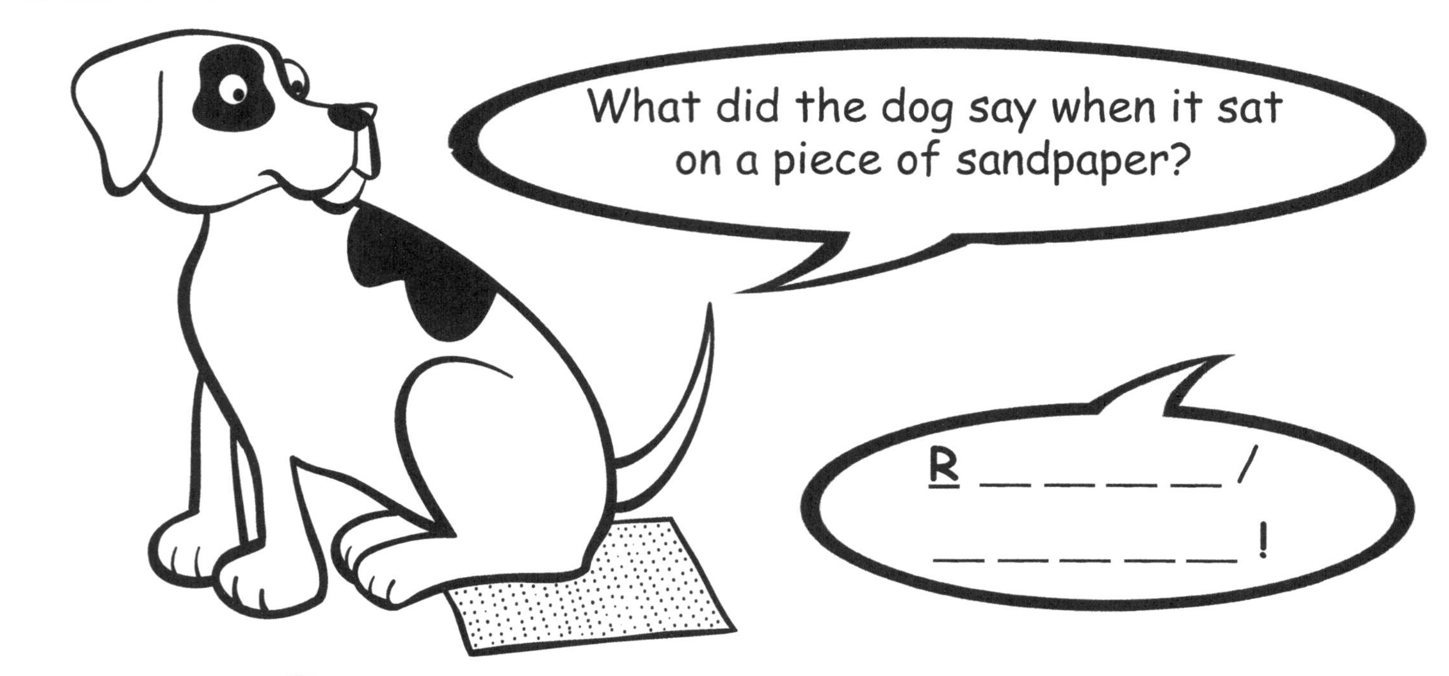

252 + 200 = **452**

205 – 7 =

394 + 60 =

979 – 50 =

998 – 900 =

412 + 40 =

208 – 10 =

654 – 200 =

129 + 800 =

118 – 20 =

Year 3 – Addition and subtraction
- *Add and subtract numbers mentally, including:*
  - *a three-digit number and ones*
  - *a three-digit number and tens*
  - *a three-digit number and hundreds.*

Add and subtract ones, tens and hundreds (4)

Learning objectives
I can add ones, tens and hundreds to a 3-digit number.
I can subtract ones, tens and hundreds from a 3-digit number.

To solve the joke, write the answer to the maths question in the oval. Then use the grid to find the letter that goes with each answer and write it on the line. The first one is done for you!

713	452	537	198	627	454	562	929	67	98
A	R	W	O	T	U	C	G	D	H

653 + 60 = **713**

497 + 40 = ( )   704 + 9 = ( )   687 – 60 = ( )

592 – 30 = ( )   298 – 200 = ( )

117 – 50 = ( )   108 + 90 = ( )   229 + 700 = ( )

Year 3 – Addition and subtraction
- *Add and subtract numbers mentally, including:*
  - *a three-digit number and ones*
  - *a three-digit number and tens*
  - *a three-digit number and hundreds.*

Add and subtract 3-digit numbers (1)

Learning objectives
I can add two 3-digit numbers together.
I can subtract one 3-digit number from another.
I can use written methods for addition and subtraction.

To solve the joke, write the answer to the maths question in the oval. Then use the grid to find the letter that goes with each answer and write it on the line. The first one is done for you!

1343	1621	748	1216	959	488	234	1075	608	243	477	801	202	579	557
U	T	S	C	P	K	L	A	E	R	G	M	I	O	D

423 + 185 = **608**

819 – 342 =

633 – 156 =

465 + 283 =

164 + 795 =

611 – 377 =

435 + 144 =

754 – 197 =

191 + 417 =

Year 3 – Addition and subtraction

- *Add and subtract numbers with up to three digits, using formal written methods of columnar addition and subtraction.*

Add and subtract 3-digit numbers (2)

Learning objectives
I can add two 3-digit numbers together.
I can subtract one 3-digit number from another.
I can use written methods for addition and subtraction.

To solve the joke, write the answer to the maths question in the oval. Then use the grid to find the letter that goes with each answer and write it on the line. The first one is done for you!

1343	1621	748	1216	959	488	234	1075	608	243	477	801	202	579	557
U	T	S	C	P	K	L	A	E	R	G	M	I	O	D

219 + 856 = **1075**

685 – 128 = ____ 582 – 339 = ____ 486 + 857 = ____

423 + 378 = ____

984 – 236 = ____ 855 + 766 = ____ 628 – 426 = ____

529 + 687 = ____ 833 – 345 = ____

Year 3 – Addition and subtraction
- *Add and subtract numbers with up to three digits, using formal written methods of columnar addition and subtraction.*

Add and subtract 3-digit numbers (3)

Learning objectives
I can add two 3-digit numbers together.
I can subtract one 3-digit number from another.
I can use written methods for addition and subtraction.

+ 3 ×
¾ = 8
1.5 ÷

To solve the joke, write the answer to the maths question in the oval. Then use the grid to find the letter that goes with each answer and write it on the line. The first one is done for you!

1252	253	1110	821	693	133	687	1403	595	1275	395
A	S	P	H	D	E	I	N	R	Y	T

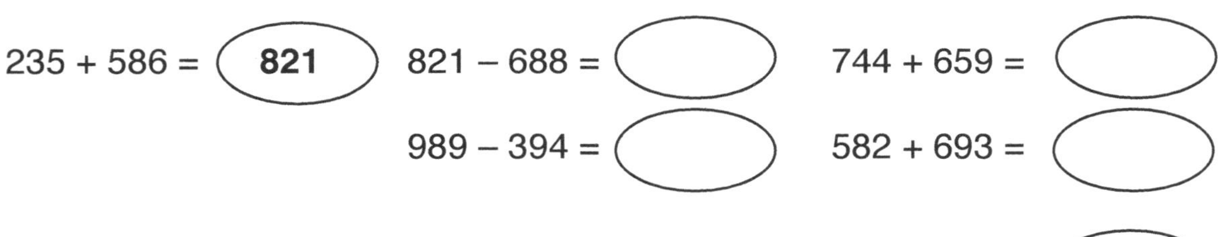

235 + 586 = **821**

821 – 688 =

989 – 394 =

744 + 659 =

582 + 693 =

654 – 259 =

566 + 255 =

657 – 524 =

564 + 688 =

862 + 248 =

738 – 605 =

Year 3 – Addition and subtraction
- *Add and subtract numbers with up to three digits, using formal written methods of columnar addition and subtraction.*

Add and subtract 3-digit numbers (4)

Learning objectives
I can add two 3-digit numbers together.
I can subtract one 3-digit number from another.
I can use written methods for addition and subtraction.

To solve the joke, write the answer to the maths question in the oval. Then use the grid to find the letter that goes with each answer and write it on the line. The first one is done for you!

1252	253	1110	821	693	133	687	1403	595	1275	395
A	S	P	H	D	E	I	N	R	Y	T

418 + 834 = **1252**

198 + 912 = ( )

577 – 444 = ( )

378 – 125 = ( )

453 + 822 = ( )

852 – 159 = ( )

867 + 385 = ( )

975 – 288 = ( )

614 – 361 = ( )

641 + 634 = ( )

Year 3 – Addition and subtraction
- *Add and subtract numbers with up to three digits, using formal written methods of columnar addition and subtraction.*

3, 4 and 8 times tables (1)

Learning objectives
I know or can work out facts for the 3, 4 and 8 times tables.
I know or can work out the related divisions for the 3, 4 and 8 times tables.

To solve the joke, write the answer to the multiplication tables question in the oval. Then use the grid to find the letter that goes with each answer and write it on the line. The first one is done for you!

12	72	96	48	1	36	5	9	15	24	7	56	8
T	I	H	A	N	E	S	R	D	U	F	L	Y

T _ _ / _ _ _ _ _ _ _ !

6 x 2 = **12** 12 x 8 = ( ) 3 x 12 = ( )

36 ÷ 4 = ( ) 3 x 8 = ( ) 8 x 7 = ( )

9 x 4 = ( ) 72 ÷ 8 = ( )

Year 3 – Multiplication and division
- *Recall and use multiplication and division facts for the 3, 4 and 8 multiplication tables.*

3, 4 and 8 times tables (2)

Learning objectives
I know or can work out facts for the 3, 4 and 8 times tables.
I know or can work out the related divisions for the 3, 4 and 8 times tables.

+ 3 ×
¾ = 8
1.5 ÷

To solve the joke, write the answer to the multiplication tables question in the oval. Then use the grid to find the letter that goes with each answer and write it on the line. The first one is done for you!

12	72	96	48	1	36	5	9	15	24	7	56	8
T	I	H	A	N	E	S	R	D	U	F	L	Y

9 x 8 = **72**

36 ÷ 3 =

96 ÷ 8=

8 x 6 =

20 ÷ 4 =

48 ÷ 4 =

12 x 3 =

5 x 3 =

56 ÷ 8=

4 x 6 =

4 ÷ 4 =

8 ÷ 8 =

32 ÷ 4 =

Year 3 – Multiplication and division
- *Recall and use multiplication and division facts for the 3, 4 and 8 multiplication tables.*

3, 4 and 8 times tables (3)

Learning objectives
I know or can work out facts for the 3, 4 and 8 times tables.
I know or can work out the related divisions for the 3, 4 and 8 times tables.

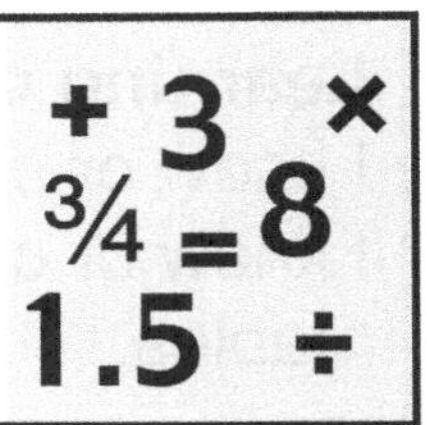

To solve the joke, write the answer to the multiplication tables question in the oval. Then use the grid to find the letter that goes with each answer and write it on the line. The first one is done for you!

11	64	3	44	1	96	33	24	48	8	16	6
C	I	E	A	S	N	Q	H	U	R	K	D

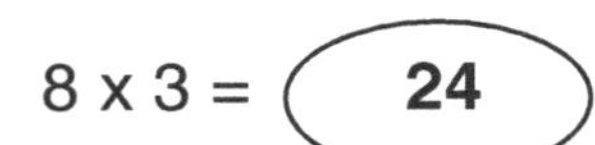
8 x 3 = 24

11 x 4 = ( )

8 x 1 = ( )

24 ÷ 4 = ( )

88 ÷ 8 = ( )

6 x 4 = ( )

9 ÷ 3 = ( )

24 ÷ 8 = ( )

4 ÷ 4 = ( )

12 ÷ 4 = ( )

Year 3 – Multiplication and division
- *Recall and use multiplication and division facts for the 3, 4 and 8 multiplication tables.*

3, 4 and 8 times tables (4)

Learning objectives
I know or can work out facts for the 3, 4 and 8 times tables.
I know or can work out the related divisions for the 3, 4 and 8 times tables.

To solve the joke, write the answer to the multiplication tables question in the oval. Then use the grid to find the letter that goes with each answer and write it on the line. The first one is done for you!

11	64	3	44	1	96	33	24	48	8	16	6
C	I	E	A	S	N	Q	H	U	R	K	D

3 x 8 = **24**　　8 x 8 = ◯　　48 ÷ 8 = ◯　　1 x 3 = ◯

4 x 11 = ◯　　12 x 8 = ◯　　18 ÷ 3 = ◯

8 ÷ 8 = ◯　　11 x 3 = ◯　　8 x 6 = ◯　　3 x 1 = ◯

11 x 4 = ◯　　4 x 4 = ◯

Year 3 – Multiplication and division
- *Recall and use multiplication and division facts for the 3, 4 and 8 multiplication tables.*

Multiply 2-digit by 1-digit numbers (1)

Learning objectives
I can multiply a 2-digit number by a 1-digit number.

To solve the joke, write the answer to the multiplication question in the oval. Then use the grid to find the letter that goes with each answer and write it on the line. The first one is done for you!

160	276	108	365	261	288	120	144	180	132	136	162
E	J	D	U	I	L	B	N	A	G	K	S

24 x 5 = **120**   45 x 4 = ( )   68 x 2 = ( )   20 x 8 = ( )

36 x 3 = ( )

15 x 8 = ( )   32 x 5 = ( )   87 x 3 = ( )   18 x 8 = ( )

44 x 3 = ( )   54 x 3 = ( )

Year 3 – Multiplication and division
- *Write and calculate mathematical statements for multiplication and division using the multiplication tables that they know, including for two-digit numbers times one-digit numbers, using mental and progressing to formal written methods.*

Multiply 2-digit by 1-digit numbers (2)

Learning objectives
I can multiply a 2-digit number by a 1-digit number.

To solve the joke, write the answer to the multiplication question in the oval. Then use the grid to find the letter that goes with each answer and write it on the line. The first one is done for you!

160	276	108	365	261	288	120	144	180	132	136	162
E	J	D	U	I	L	B	N	A	G	K	S

60 x 3 = 180

69 x 4 = ( )   73 x 5 = ( )   48 x 3 = ( )   33 x 4 = ( )

36 x 8 = ( )   80 x 2 = ( )

81 x 2 = ( )   60 x 3 = ( )   96 x 3 = ( )   40 x 4 = ( )

Year 3 – Multiplication and division

- *Write and calculate mathematical statements for multiplication and division using the multiplication tables that they know, including for two-digit numbers times one-digit numbers, using mental and progressing to formal written methods*

Multiply 2-digit by 1-digit numbers (3)

Learning objectives
I can multiply a 2-digit number by a 1-digit number.

To solve the joke, write the answer to the multiplication question in the oval. Then use the grid to find the letter that goes with each answer and write it on the line. The first one is done for you!

252	168	204	156	280	128	240	188	200	395	136	260
B	D	R	O	I	L	E	Y	Q	W	U	A

56 x 3 = (168) 39 x 4 = ( ) 32 x 4 = ( ) 16 x 8 = ( )

94 x 2 = ( )

79 x 5 = ( ) 78 x 2 = ( ) 52 x 3 = ( ) 21 x 8 = ( )

Year 3 – Multiplication and division

- *Write and calculate mathematical statements for multiplication and division using the multiplication tables that they know, including for two-digit numbers times one-digit numbers, using mental and progressing to formal written methods*

Multiply 2-digit by 1-digit numbers (4)

Learning objectives
I can multiply a 2-digit number by a 1-digit number.

+ 3 ×
¾ = 8
1.5 ÷

To solve the joke, write the answer to the multiplication question in the oval. Then use the grid to find the letter that goes with each answer and write it on the line. The first one is done for you!

252	168	204	156	280	128	240	188	200	395	136	260
B	D	R	O	I	L	E	Y	Q	W	U	A

52 x 5 = **260**

84 x 3 = ( )

65 x 4 = ( )

68 x 3 = ( )

63 x 4 = ( )

35 x 8 = ( )

30 x 8 = ( )

25 x 8 = ( )

34 x 4 = ( )

60 x 4 = ( )

17 x 8 = ( )

80 x 3 = ( )

Year 3 – Multiplication and division

- *Write and calculate mathematical statements for multiplication and division using the multiplication tables that they know, including for two-digit numbers times one-digit numbers, using mental and progressing to formal written methods.*

Recognise and find fractions (1)

Learning objectives
I can write how much of a shape is shaded as a fraction.

To solve the joke, work out what fraction of the shape is shaded. Write the answer in the oval then use the grid to find the letter that goes with each answer and write it on the line. The first one is done for you!

1/2	3/8	5/6	3/5	2/3	4/9	2/5
S	A	N	P	O	E	R

Year 3 – Fractions
- *Recognise, find and write fractions of a discrete set of objects: unit fractions and non-unit fractions with small denominators.*

Recognise and find fractions (2)

Learning objectives
I can find fractions of a number or set of objects.

To solve the joke, work out the fraction question. Write the answer in the circle then use the grid to find the letter that goes with each answer and write it on the line. The first one is done for you!

8	7	5	3	6	4	2	1	9
I	A	C	S	P	O	E	T	R

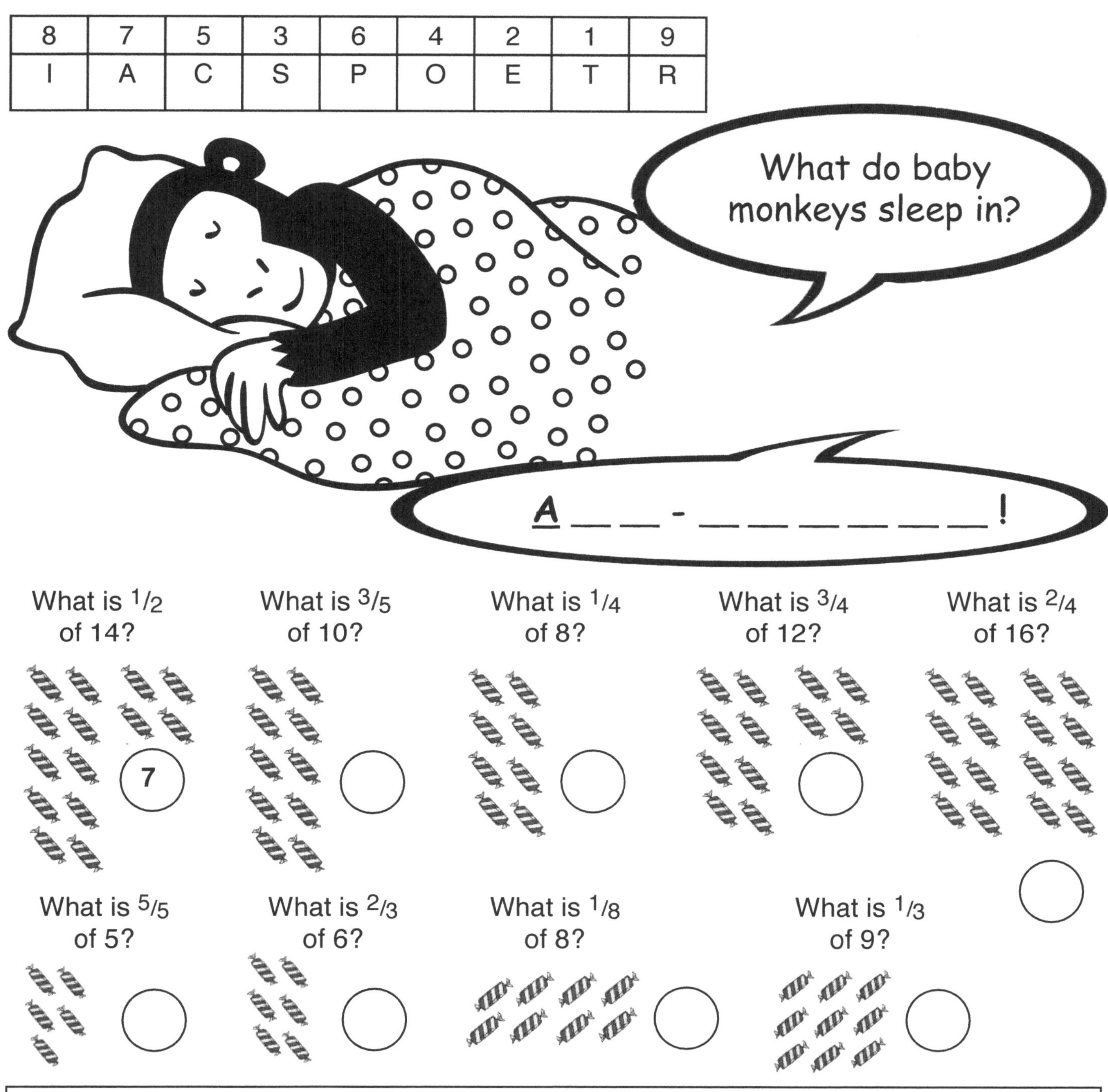

Year 3 – Fractions

- *Recognise, find and write fractions of a discrete set of objects: unit fractions and non-unit fractions with small denominators.*

Recognise and find fractions (3)

Learning objectives
I can write how much of a shape is shaded as a fraction.

To solve the joke, work out what fraction of the shape is shaded. Then use the grid to find the letter that goes with each answer and write it on the line. The first one is done for you!

1/3	3/4	5/8	5/6	3/8
C	O	M	S	I

Year 3 – Fractions

- Recognise, find and write fractions of a discrete set of objects: unit fractions and non-unit fractions with small denominators.

Recognise and find fractions (4)

Learning objectives
I can find fractions of a number or set of objects.

+ 3 ×
¾ = 8
1.5 ÷

To solve the joke, work out the fraction question. Write the answer in the oval then use the grid to find the letter that goes with each answer and write it on the line. The first one is done for you!

1	4	2	5	3	6
S	T	O	H	M	E

What is $2/3$ of 6? **4**

What is $1/4$ of 20?

What is $2/4$ of 12?

What is $3/3$ of 3?

What is $1/5$ of 10?

What is $2/8$ of 8?

What is $1/7$ of 7?

Year 3 – Fractions
- *Recognise, find and write fractions of a discrete set of objects: unit fractions and non-unit fractions with small denominators.*

Add and subtract fractions (1)

Learning objectives
I can add two fractions together.
I can subtract one fraction from another.

To solve the joke, work out the answer to the fractions question. Write the answer in the circle, then use the grid to find the letter that goes with each answer and write it on the line. The first one is done for you!

$2/5$	$5/8$	$7/9$	$4/9$	$6/8$	$5/6$	$5/5$	$3/7$	$6/9$	$6/7$	$7/8$	$2/6$
E	R	S	B	Q	L	U	I	Y	T	H	D

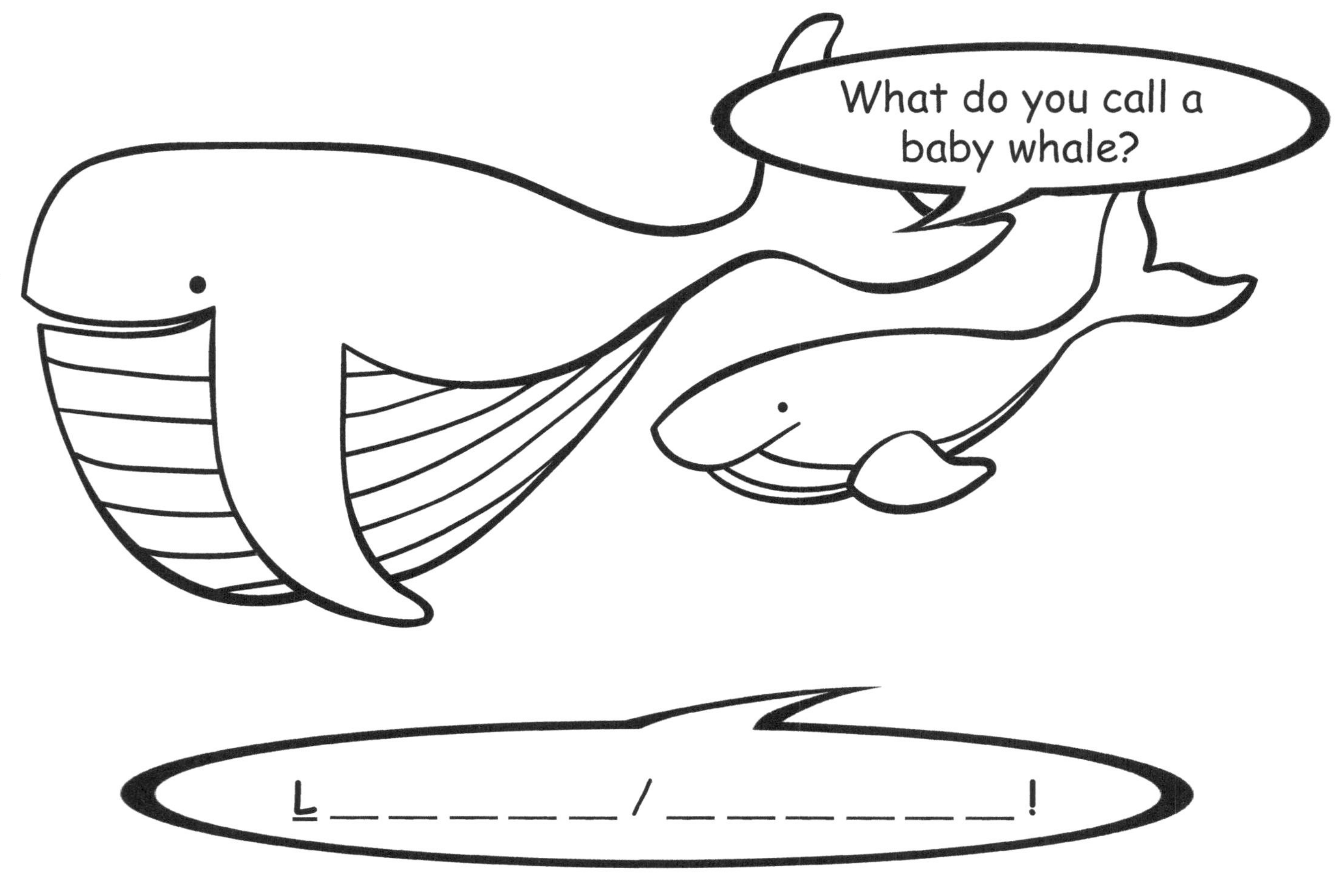

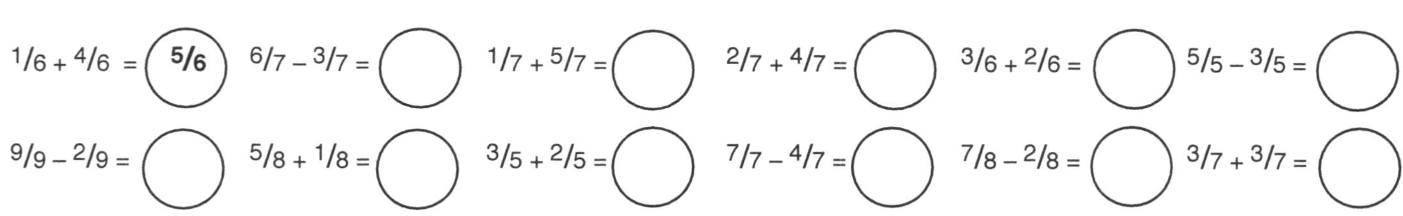

$1/6 + 4/6 =$ (5/6) $6/7 - 3/7 =$ ( ) $1/7 + 5/7 =$ ( ) $2/7 + 4/7 =$ ( ) $3/6 + 2/6 =$ ( ) $5/5 - 3/5 =$ ( )

$9/9 - 2/9 =$ ( ) $5/8 + 1/8 =$ ( ) $3/5 + 2/5 =$ ( ) $7/7 - 4/7 =$ ( ) $7/8 - 2/8 =$ ( ) $3/7 + 3/7 =$ ( )

Year 3 – Fractions
- *Add and subtract fractions with the same denominator within one whole (for example, $5/7 + 1/7 = 6/7$).*

Add and subtract fractions (2)

Learning objectives
I can add two fractions together.
I can subtract one fraction from another.

+ 3 ×
¾ = 8
1.5 ÷

To solve the joke, work out the answer to the fractions question. Write the answer in the circle, then use the grid to find the letter that goes with each answer and write it on the line. The first one is done for you!

$^2/_5$	$^5/_8$	$^7/_9$	$^4/_9$	$^6/_8$	$^5/_6$	$^5/_5$	$^3/_7$	$^6/_9$	$^6/_7$	$^7/_8$	$^2/_6$
E	R	S	B	Q	L	U	I	Y	T	H	D

B _ _ _ _ _ / _ _ _ _ / _ _ _ _ _ _ _ !

$^8/_9 - ^4/_9 =$ (4/9) $^1/_7 + ^2/_7 =$ ( ) $^6/_6 - ^1/_6 =$ ( ) $^2/_6 + ^3/_6 =$ ( ) $^8/_9 - ^2/_9 =$ ( )

$^7/_7 - ^1/_7 =$ ( ) $^2/_8 + ^5/_8 =$ ( ) $^1/_5 + ^1/_5 =$ ( )

$^3/_9 + ^4/_9 =$ ( ) $^3/_8 + ^3/_8 =$ ( ) $^1/_5 + ^4/_5 =$ ( ) $^5/_7 - ^2/_7 =$ ( ) $^5/_6 - ^3/_6 =$ ( )

Year 3 – Fractions

- *Add and subtract fractions with the same denominator within one whole (for example, $^5/_7 + ^1/_7 = ^6/_7$).*

Add and subtract fractions (3)

Learning objectives
I can add two fractions together.
I can subtract one fraction from another.

To solve the joke, work out the answer to the fractions question. Write the answer in the circle, then use the grid to find the letter that goes with each answer and write it on the line. The first one is done for you!

$2/5$	$5/8$	$7/9$	$4/9$	$6/8$	$5/6$	$5/5$	$3/7$	$6/9$	$6/7$	$1/5$	$2/6$	$1/4$	$7/8$
P	R	L	O	A	Y	C	S	B	K	N	U	H	E

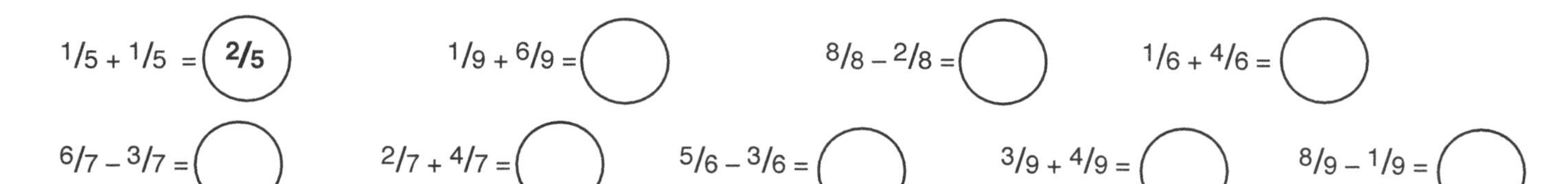

$1/5 + 1/5 =$ **2/5** $1/9 + 6/9 =$ ◯ $8/8 - 2/8 =$ ◯ $1/6 + 4/6 =$ ◯

$6/7 - 3/7 =$ ◯ $2/7 + 4/7 =$ ◯ $5/6 - 3/6 =$ ◯ $3/9 + 4/9 =$ ◯ $8/9 - 1/9 =$ ◯

Year 3 – Fractions
- *Add and subtract fractions with the same denominator within one whole (for example, $5/7 + 1/7 = 6/7$)*

Add and subtract fractions (4)

Learning objectives
I can add two fractions together.
I can subtract one fraction from another.

To solve the joke, work out the answer to the fractions question. Write the answer in the circle, then use the grid to find the letter that goes with each answer and write it on the line. The first one is done for you!

$2/5$	$5/8$	$7/9$	$4/9$	$6/8$	$5/6$	$5/5$	$3/7$	$6/9$	$6/7$	$1/8$	$2/6$	$1/4$	$7/8$
P	R	L	O	A	Y	C	S	B	K	N	U	H	E

$1/7 + 2/7 =$ ($3/7$) $3/4 - 2/4 =$ ( ) $5/8 + 2/8 =$ ( ) $7/8 - 2/8 =$ ( )

$9/9 - 2/9 =$ ( ) $2/9 + 2/9 =$ ( ) $1/5 + 4/5 =$ ( ) $3/7 + 3/7 =$ ( )

$7/9 - 1/9 =$ ( ) $6/9 - 2/9 =$ ( ) $5/8 - 4/8 =$ ( ) $4/8 + 3/8 =$ ( ) $7/7 - 4/7 =$ ( )

Year 3 – Fractions
- *Add and subtract fractions with the same denominator within one whole (for example,* $5/7 + 1/7 = 6/7$*)*

Compare and order fractions (1)

Learning objectives

I can compare fractions with a numerator of 1 and say which is the largest.

I can compare fractions with the same denominator and say which is the largest.

To solve the joke, work out which is the largest fraction. Write the answer in the circle, then use the grid to find the letter that goes with each answer and write it on the line. The first one is done for you!

1/5	3/4	2/3	4/5	1/1	3/7	3/5	6/7	5/8	1/3	1/4	1/2	8/9	4/7
Y	A	B	L	I	E	N	T	D	C	M	S	H	R

3/4 2/4 1/4 (**3/4**)   3/5 4/5 1/5 ( )   2/5 1/5 4/5 ( )   3/7 1/7 2/7 ( )   1/7 1/5 1/8 ( )

1/5 1/4 1/3 ( )   2/4 1/4 3/4 ( )   4/7 2/7 6/7 ( )   1/5 1/2 1/3 ( )

Year 3 – Fractions

- *Compare and order unit fractions, and fractions with the same denominators.*

Compare and order fractions (2)

Learning objectives
I can compare fractions with a numerator of 1 and say which is the largest.
I can compare fractions with the same denominator and say which is the largest.

+ 3 ×
¾ = 8
1.5 ÷

To solve the joke, work out which is the largest fraction. Write the answer in the circle, then use the grid to find the letter that goes with each answer and write it on the line. The first one is done for you!

1/5	3/4	2/3	4/5	1/1	3/7	3/5	6/7	5/8	1/3	1/4	1/2	8/9	4/7
Y	A	B	L	I	E	N	T	D	C	M	S	H	R

1/7 3/7 6/7 (6/7) 5/9 2/9 8/9 ( ) 2/7 4/7 3/7 ( ) 3/7 1/7 2/7 ( ) 1/7 3/7 2/7 ( )

1/3 2/3 0/3 ( ) 2/5 1/5 4/5 ( ) 1/1 1/5 1/2 ( ) 2/5 3/5 1/5 ( ) 3/8 5/8 4/8 ( )

1/4 1/5 1/6 ( ) 1/4 1/1 1/3 ( ) 1/3 1/4 1/5 ( ) 1/7 3/7 0/7 ( )

Year 3 – Fractions
- *Compare and order unit fractions, and fractions with the same denominators.*

Compare and order fractions (3)

Learning objectives
I can compare fractions with a numerator of 1 and say which is the largest.
I can compare fractions with the same denominator and say which is the largest.

To solve the joke, work out which is the largest fraction. Write the answer in the circle, then use the grid to find the letter that goes with each answer and write it on the line. The first one has been done for you!

3/4	4/5	6/7	3/5	1/6	5/8	1/4	7/9	8/9	1/3	7/7	4/7
R	O	A	D	P	K	I	H	E	T	S	C

3/5 2/5 1/5 (**3/5**)   1/8 1/4 1/5 ( )   2/9 8/9 6/9 ( )   1/3 1/4 1/6 ( )

4/7 2/7 3/7 ( )   3/4 2/4 1/4 ( )   2/5 4/5 3/5 ( )   3/7 5/7 6/7 ( )   4/8 5/8 2/8 ( )

Year 3 – Fractions
- *Compare and order unit fractions, and fractions with the same denominators.*

Compare and order fractions (4)

Learning objectives

I can compare fractions with a numerator of 1 and say which is the largest.

I can compare fractions with the same denominator and say which is the largest.

To solve the joke, work out which is the largest fraction. Write the answer in the circle, then use the grid to find the letter that goes with each answer and write it on the line. The first one has been done for you!

3/4	4/5	6/7	3/5	1/6	5/8	1/4	7/9	8/9	1/3	7/7	4/7
R	O	A	D	P	K	I	H	E	T	S	C

7/9 3/9 1/9 (7/9)   1/5 2/5 4/5 ( )   1/4 1/6 1/3 ( )

4/7 1/7 3/7 ( )   3/4 2/4 1/4 ( )   1/5 4/5 2/5 ( )   1/7 2/7 6/7 ( )   1/8 5/8 3/8 ( )

3/5 1/5 4/5 ( )   5/7 6/7 4/7 ( )

Year 3 – Fractions

- *Compare and order unit fractions, and fractions with the same denominators.*

Sequence multiples of 6, 7, 9, 25 and 1000 (1)

Learning objectives
I can count forwards and backwards in multiples of 6, 7, 9, 25 and 1000.

+ 3 ×
¾ = 8
1.5 ÷

To solve the jokes, work out the number that comes next and write it in the oval. Then use the grid to find the letter that goes with each answer and write it on the line. The first one is done for you!

72	90	49	1000	108	275	54	5000	84
A	I	D	T	M	G	E	S	W

What happens to a green stone if you throw it in the red sea?

72, 78, 84, **90**

4000, 3000, 2000, ( )

I __ / __ __ __ __ __ / __ __ __ __!

350, 325, 300, ( )

72, 66, 60, ( )

925, 950, 975, ( )

8000, 7000, 6000, ( )

63, 70, 77, ( )

27, 36, 45, ( )

1075, 1050, 1025, ( )

What cheese is made backwards?

__ __ __ __ __!

81, 72, 63, ( )

70, 63, 56, ( )

54, 60, 66, ( )

81, 90, 99, ( )

Year 4 – Number and place value
- *Count in multiples of 6, 7, 9, 25 and 1000.*

Sequence multiples of 6, 7, 9, 25 and 1000 (2)

Learning objectives
I can count forwards and backwards in multiples of 6, 7, 9, 25 and 1000.

+ 3 ×
¾ = 8
1.5 ÷

To solve the jokes, work out the number that comes next and write it in the oval. Then use the grid to find the letter that goes with each answer and write it on the line. The first one is done for you!

24	81	56	10 000	90	375	72	7000	42	650
T	I	H	S	E	P	M	A	O	V

Where do cows go for a night out?

42, 36, 30, **24** 35, 42, 49, 63, 72, 81,

54, 60, 66, 63, 56, 49, 24, 30, 36,

575, 600, 625, 108, 99, 90, 72, 78, 84, 7000, 8000, 9000,

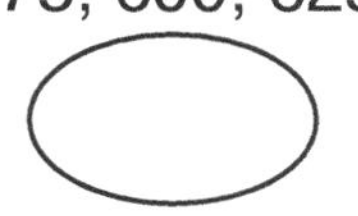

What do you call a man with cow poo on his head?

_ _ _ !

450, 425, 400, 10000, 9000, 8000, 6, 12, 18,

Year 4 – Number and place value
- *Count in multiples of 6, 7, 9, 25 and 1000.*

1000 more or less (1)

Learning objectives
I can find 1000 more than a number.
I can find 1000 less than a number.

To solve the joke, work out the answer to the maths question and write it in the oval. Then use the grid to find the letter that goes with each answer and write it on the line. The first one is done for you!

2098	5247	3606	4087	5606	4906	3547	5547	2398
B	T	A	H	L	E	Y	M	O

What is 1000 more than 4247? **5247**

What is 1000 less than 5087? ( )

What is 1000 more than 3906? ( )

What is 1000 less than 6547? ( )

What is 1000 less than 3398? ( )

What is 1000 less than 6247? ( )

What is 1000 more than 3087? ( )

What is 1000 less than 3098? ( )

What is 1000 more than 2606? ( )

What is 1000 more than 4606? ( )

What is 1000 less than 6606? ( )

Year 4 – Number and place value
- *Find 1000 more or less than a given number.*

1000 more or less (2)

Learning objectives
I can find 1000 more than a number.
I can find 1000 less than a number.

$+\ 3\ \times$
$\frac{3}{4} = 8$
$1.5\ \div$

To solve the jokes, work out the answer to the maths question and write it in the oval. Then use the grid to find the letter that goes with each answer and write it on the line. The first one is done for you!

2098	5247	3606	4087	5606	4906	3547	5547	2398
B	T	A	H	L	E	Y	M	O

What do bees say when they can't decide?

What is 1000 more than 4547? **5547**

What is 1000 less than 4606?

What is 1000 more than 2547?

What is 1000 more than 1098?

What is 1000 less than 5906?

What is 1000 more than 3906?

3698	8649	5720	897	9511	4856	6747	512	247	3578	8524	1919
L	E	O	S	N	A	G	W	T	V	I	M

What do the Atlantic and Pacific oceans do when they greet each other?

____ !

What is 1000 less than 1512?

What is 1000 more than 3856?

What is 1000 less than 4578?

What is 1000 more than 7649?

Year 4 – Number and place value
- *Find 1000 more or less than a given number.*

1000 more or less (3)

Learning objectives
I can find 1000 more than a number.
I can find 1000 less than a number.

To solve the joke, work out the answer to the maths question and write it in the oval. Then use the grid to find the letter that goes with each answer and write it on the line. The first one is done for you!

3698	8649	5720	897	9511	4856	6747	512	247	3578	8524	1919
L	E	O	S	N	A	G	W	T	V	I	M

What is 1000 less than 4698? **3698**

What is 1000 less than 6720?

What is 1000 more than 8511?

What is 1000 more than 5747?

What is 1000 less than 1247?

What is 1000 less than 9524?

What is 1000 more than 919?

What is 1000 less than 9649?

What is 1000 more than 8511?

What is 1000 more than 4720?

What is 1000 less than 1897?

What is 1000 more than 7649?

What is 1000 less than 5856?

Year 4 – Number and place value
- *Find 1000 more or less than a given number.*

Counting through negative numbers (1)

Learning objectives
I can count backwards through zero into negative numbers.

+ 3 ×
¾ = 8
1.5 ÷

To solve the joke, work out the number that should go in the circle. Then use the grid to find the letter that goes with each answer and write it on the line. The first one is done for you.

-1	-5	-13	-2	-9	-3	-10	0	-6	-11	-12	-8	-4	-7
H	L	D	N	R	E	O	G	P	F	I	S	T	A

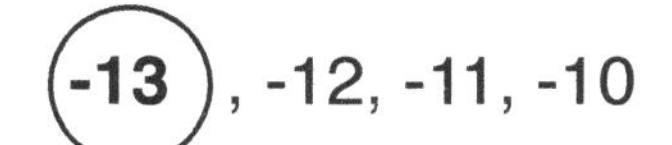

(-13), -12, -11, -10 ( ), -8, -7, -6 ( ), -9, -8, -7 ( ), -5, -4, -3

( ), -11, -10, -9 ( ), -3, -2, -1 ( ), -6, -5, -4

( ), -4, -3, -2 ( ), -11, -10, -9 ( ), -1, 0, 1 ( ), -2, -1, 0

Year 4 – Number and place value
- *Count backwards through zero to include negative numbers.*

Counting through negative numbers (2)

Learning objectives
I can count backwards through zero into negative numbers.

To solve the joke, work out the number that should go in the circle. Then use the grid to find the letter that goes with each answer and write it on the line. The first one is done for you!

-1	-5	-13	-2	-9	-3	-10	0	-6	-11	-12	-8	-4	-7
H	L	D	N	R	E	O	G	P	F	I	S	T	A

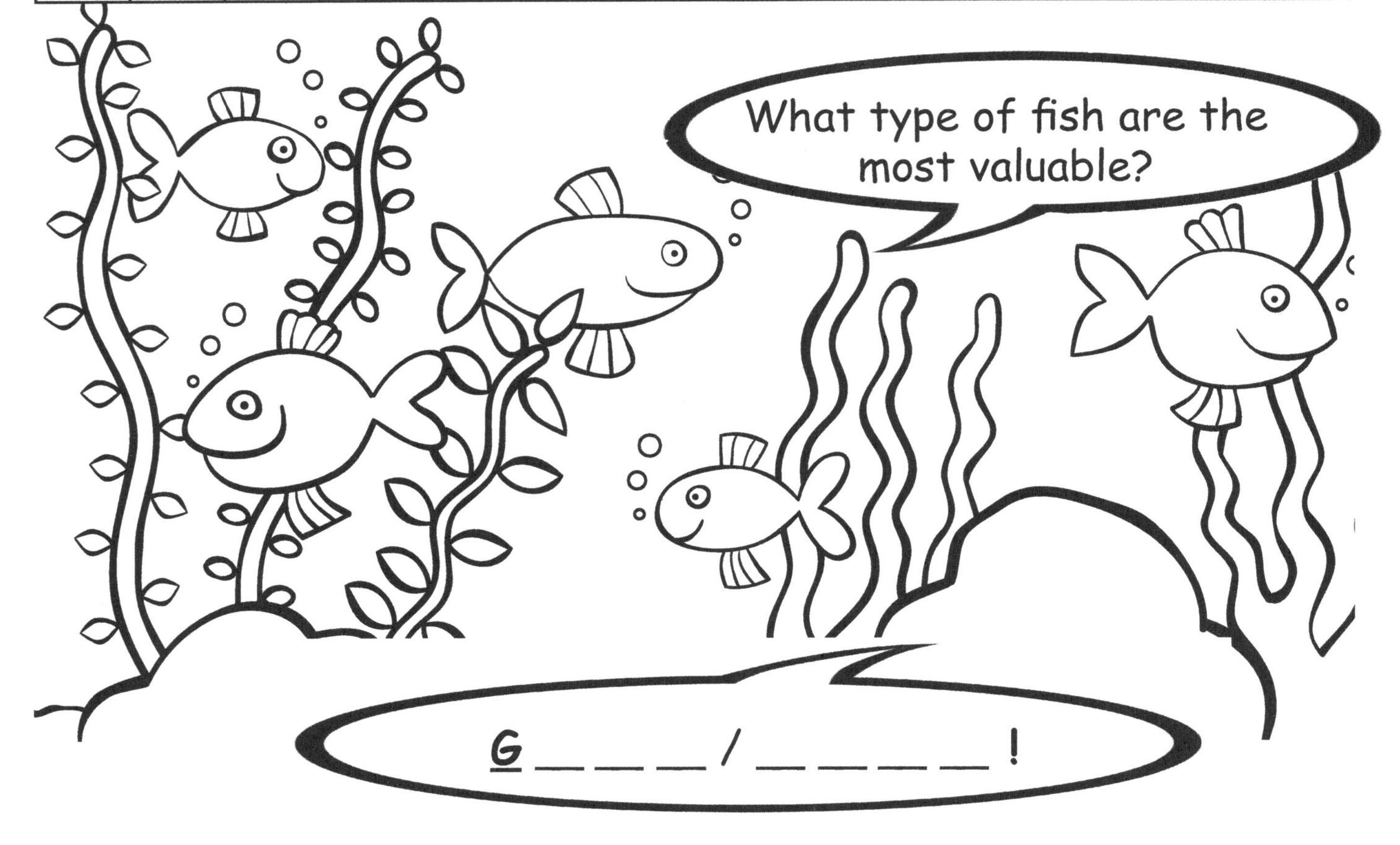

G _ _ _ _ / _ _ _ _ _ _ !

(0), 1, 2, 3　　( ), -9, -8, -7　　( ), -4, -3, -2　　( ), -12, -11, 10

( ), -10, -9, -8　　( ), -11, -10, -9　　( ), -7, -6, -5　　( ), 0, 1, 2

Year 4 – Number and place value
- *Count backwards through zero to include negative numbers.*

Counting through negative numbers (3)

Learning objectives
I can count backwards through zero into negative numbers.

To solve the joke, work out the number that should go in the circle. Then use the grid to find the letter that goes with each answer and write it on the line. The first one is done for you.

-1	-5	-8	-2	-9	-3	-10	0	-6	-7	-4
I	S	T	G	E	R	L	A	M	V	F

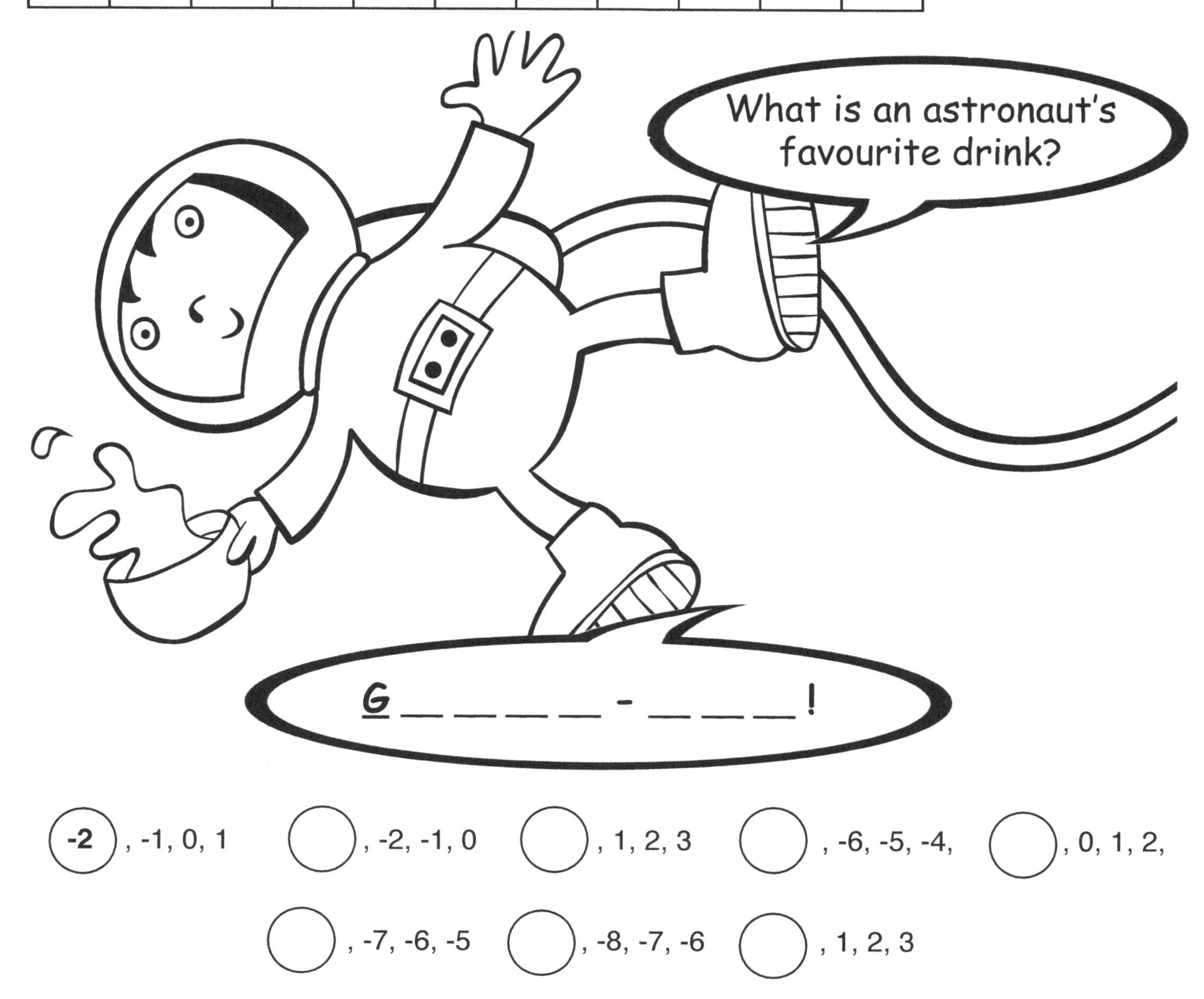

(-2), -1, 0, 1 ( ), -2, -1, 0 ( ), 1, 2, 3 ( ), -6, -5, -4, ( ), 0, 1, 2,

( ), -7, -6, -5 ( ), -8, -7, -6 ( ), 1, 2, 3

Year 4 – Number and place value
- *Count backwards through zero to include negative numbers.*

Counting through negative numbers (4)

Learning objectives
I can count backwards through zero into negative numbers.

To solve the joke, work out the number that should go in the circle. Then use the grid to find the letter that goes with each answer and write it on the line. The first one is done for you!

-1	-5	-8	-2	-9	-3	-10	0	-6	-7	-4
I	S	T	G	E	R	L	A	M	V	F

(-5), -4, -3, -2

( ), -9, -8, -7,

( ), 0, 1, 2

( ), -5, -4, -3

( ), -8, -7, -6

( ), -3, -2, -1

( ), -9, -8, -7

( ), 0, 1, 2

( ), -8, -7, -6

( ), -4, -3, -2

Year 4 – Number and place value
- *Count backwards through zero to include negative numbers.*

Place value Th, H, T, O (1)

Learning objectives
I know the value of each digit in numbers up to 10 000.

+ 3 ×
¾ = 8
1.5 ÷

To solve the joke, use place value to work out the value of the underlined digit and write the answer in the oval. Then use the grid to find the letter that goes with each answer and write it on the line. The first one is done for you!

9000	3000	200	4000	30	8	80	1	400	900	40	7000	60	700
C	E	S	D	L	H	I	A	F	M	W	P	O	R

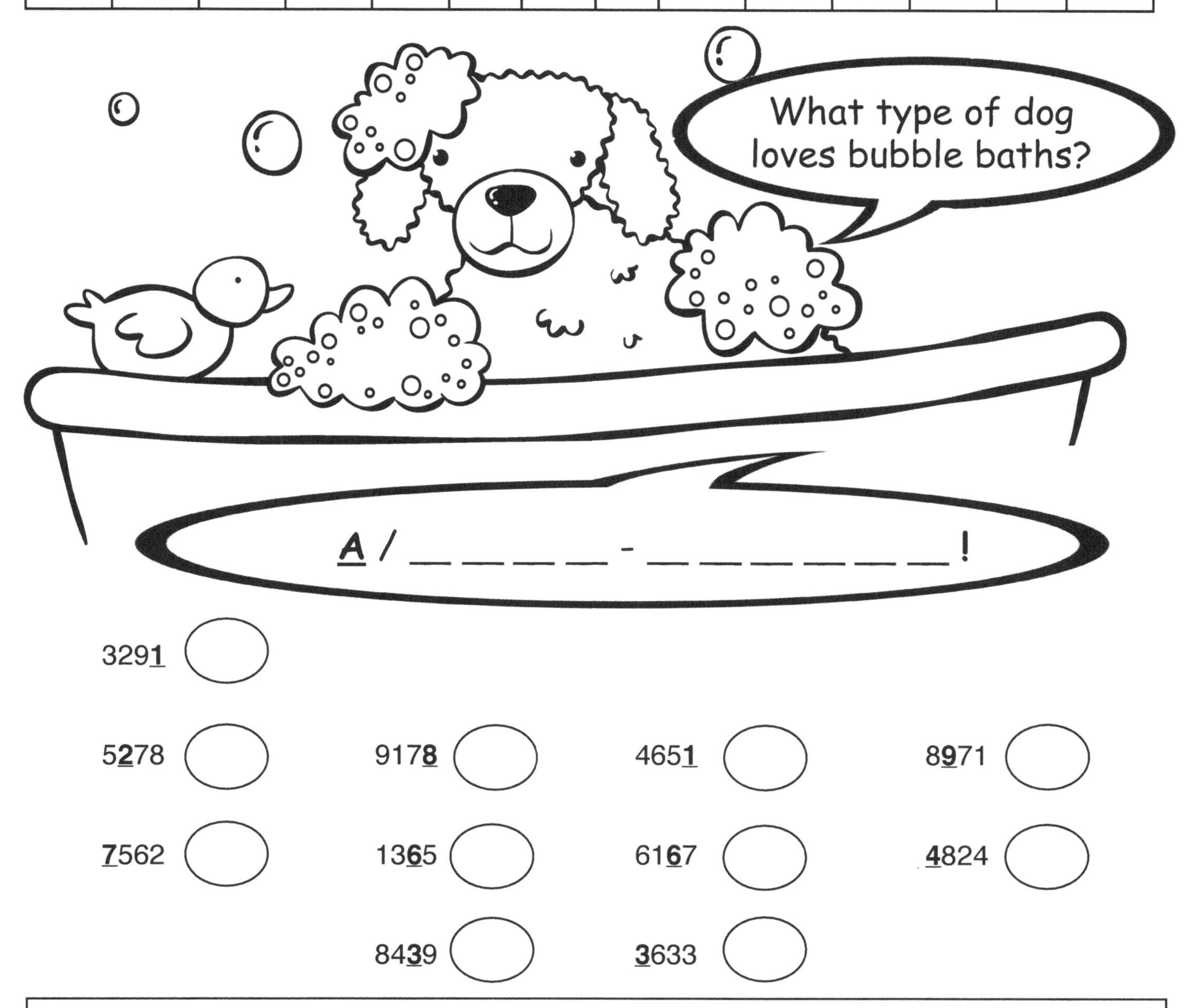

Year 4 – Number and place value
- *Recognise the place value of each digit in a four-digit number (thousands, hundreds, tens, and ones).*

Place value Th, H, T, O (2)

Learning objectives
I know the value of each digit in numbers up to 10 000.

To solve the joke, use place value to work out the value of the underlined digit and write the answer in the oval. Then use the grid to find the letter that goes with each answer and write it on the line. The first one is done for you!

9000	3000	200	4000	30	8	80	1	400	900	40	7000	60	700
C	E	S	D	L	H	I	A	F	M	W	P	O	R

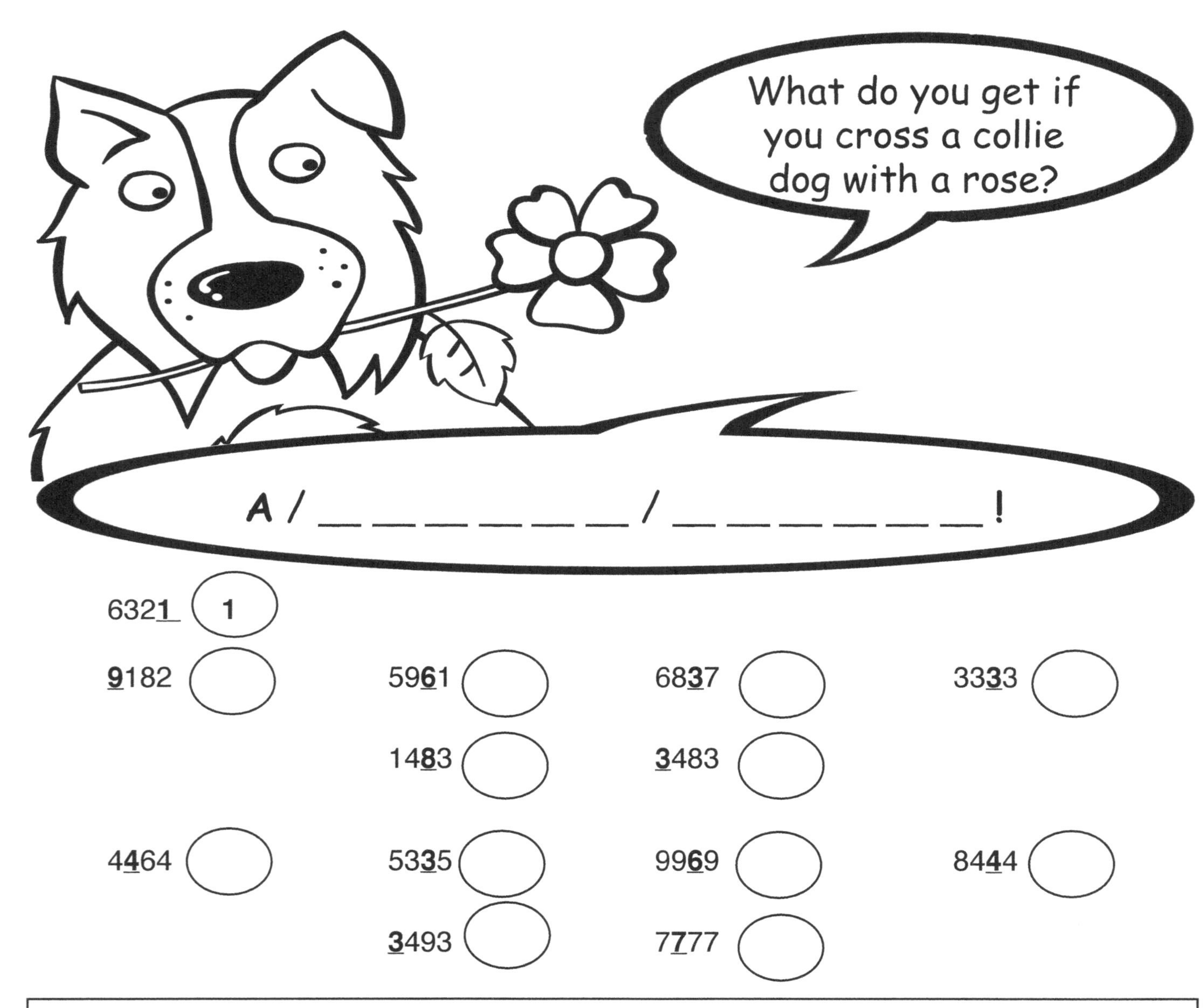

632**1** (1)

9182

59**6**1

68**3**7

33**3**3

14**8**3

3483

4**4**64

53**3**5

99**6**9

84**4**4

3493

7**7**77

Year 4 – Number and place value

- *Recognise the place value of each digit in a four-digit number (thousands, hundreds, tens, and ones).*

Place value Th, H, T, O (3)

Learning objectives
I know the value of each digit in numbers up to 10 000.

To solve the joke, use place value to work out the value of the underlined digit and write the answer in the oval. Then use the grid to find the letter that goes with each answer and write it on the line. The first one is done for you!

700	5000	200	2000	50	7	70	5	500	900	20	9000
S	L	T	R	A	B	F	O	E	P	H	I

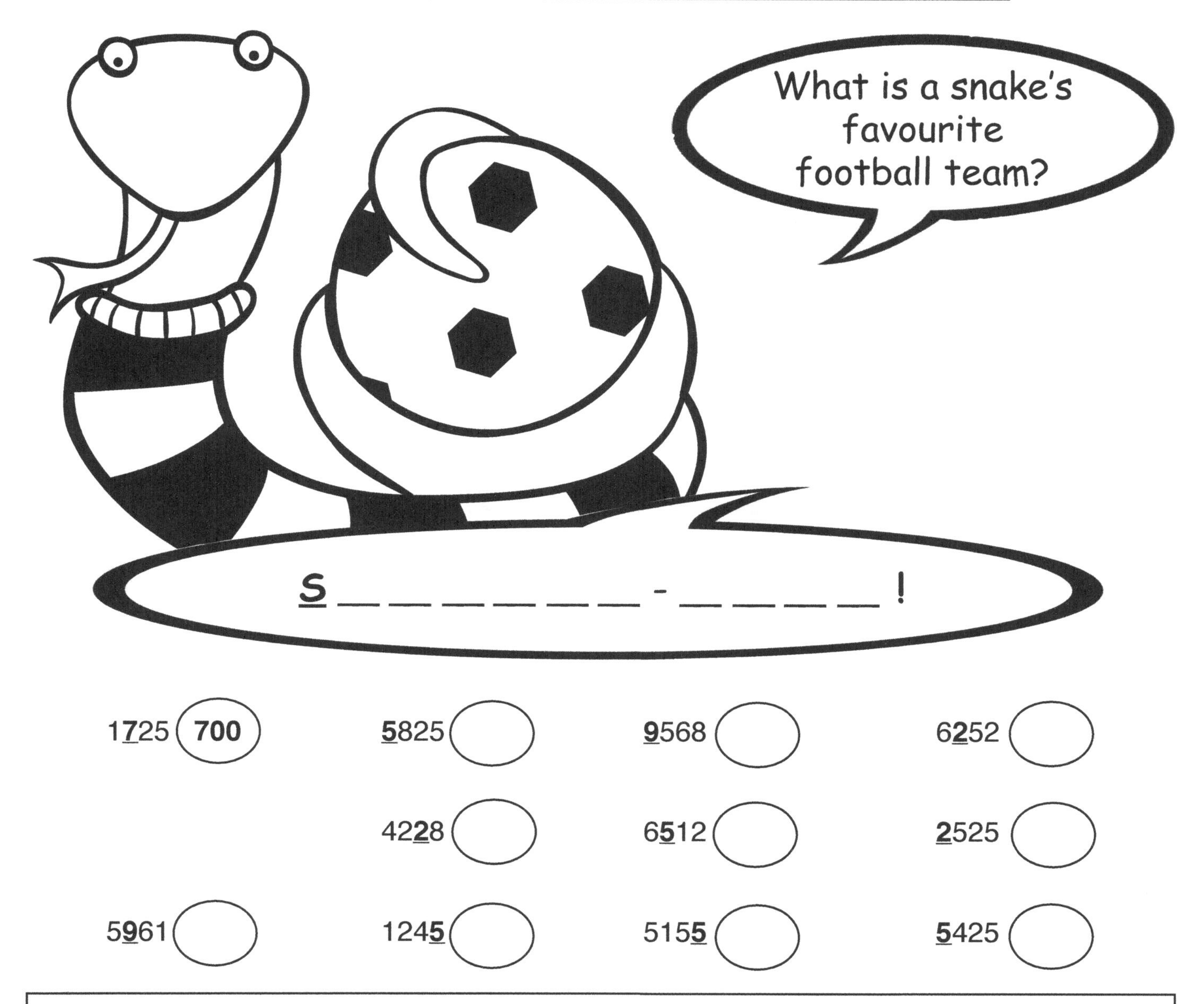

1**7**25 (700) | **5**825 ( ) | **9**568 ( ) | 6**2**52 ( )

4**2**28 ( ) | 6**5**12 ( ) | **2**525 ( )

5**9**61 ( ) | 124**5** ( ) | 515**5** ( ) | **5**425 ( )

Year 4 –Number and place value
- *Recognise the place value of each digit in a four-digit number (thousands, hundreds, tens, and ones).*

Place value Th, H, T, O (4)

Learning objectives
I know the value of each digit in numbers up to 10 000

To solve the joke, use place value to work out the value of the underlined digit and write the answer in the oval. Then use the grid to find the letter that goes with each answer and write it on the line. The first one is done for you!

700	5000	200	2000	50	7	70	5	500	900	20	9000
S	L	T	R	A	B	F	O	E	P	H	I

14**5**8 (50)

77**7**7 ( ) 1**5**45 ( ) 55**5**5 ( ) 3**2**26 ( )

22**2**5 ( ) 3**5**65 ( ) **2**228 ( )

174**7** ( ) 555**5** ( ) 46**5**5 ( )

Year 4 –Number and place value
- *Recognise the place value of each digit in a four-digit number (thousands, hundreds, tens, and ones).*

Rounding to 10, 100 or 1000 (1)

Learning objectives
I can round numbers to the nearest 10, 100 or 1000.

To solve the joke, round each number and write the answer in the oval. Then use the grid to find the letter that goes with each answer and write it on the line. The first one is done for you!

6000	5200	570	400	990	2300	180	7000	100	8000
S	O	U	N	E	A	W	C	T	B

What is 5245 to the nearest 100? **5200**

What is 356 to the nearest 100?

What is 2289 to the nearest 100?

What is 419 to the nearest 100?

What is 5228 to the nearest 100?

What is 7445 to the nearest 1000?

What is 148 to the nearest 100?

What is 5151 to the nearest 100?

What is 7811 to the nearest 1000?

What is 567 to the nearest 10?

What is 6299 to the nearest 1000?

Year 4 – Number and place value

- *Round any number to the nearest 10, 100 or 1000.*

Rounding to 10, 100 or 1000 (2)

Learning objectives
I can round numbers to the nearest 10, 100 or 1000.

To solve the joke, round each number and write the answer in the circle. Then use the grid to find the letter that goes with each answer and write it on the line. The first one is done for you!

6000	5200	570	400	990	2300	180	7000	100	8000
S	O	U	N	E	A	W	C	T	B

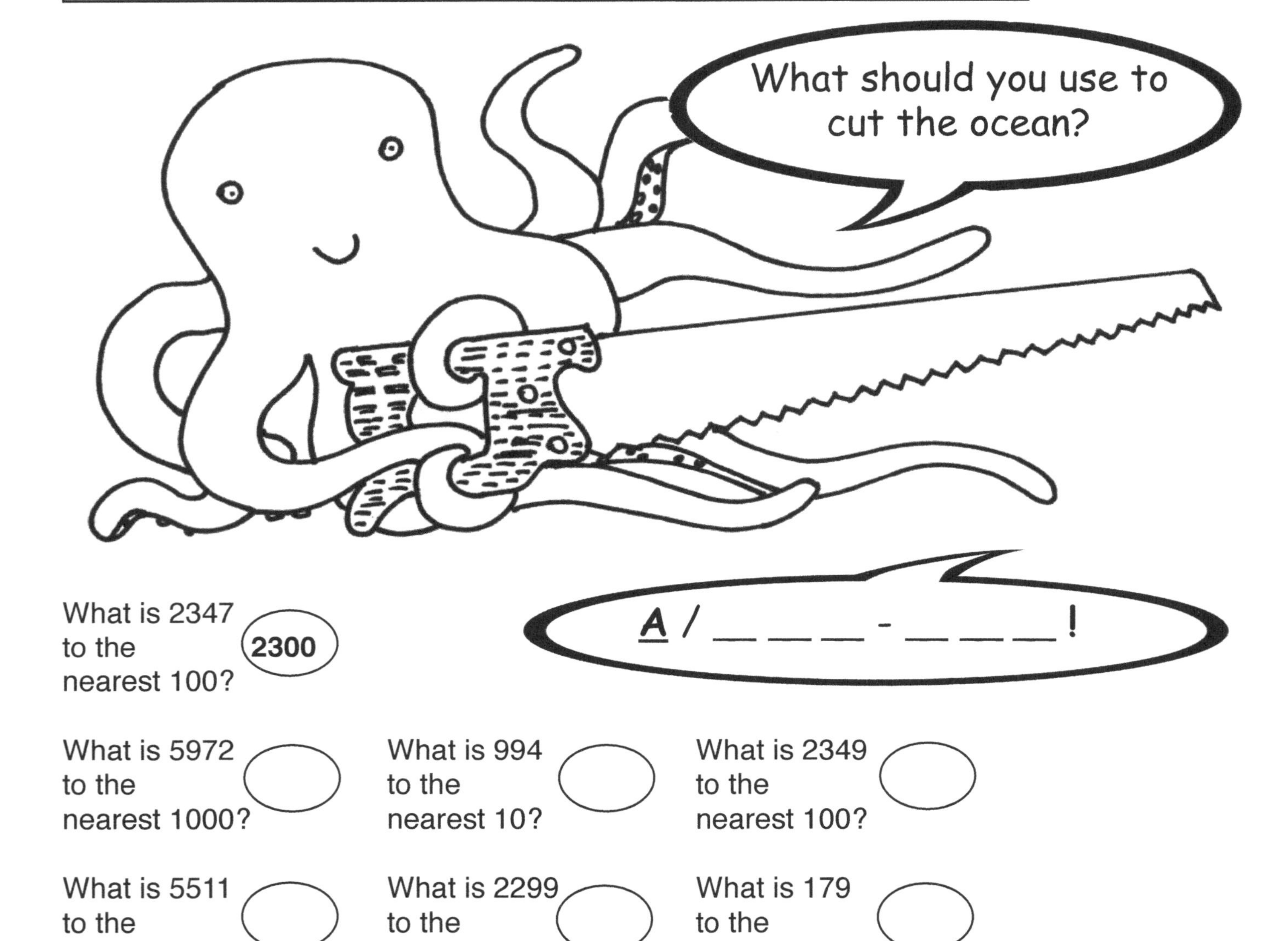

What is 2347 to the nearest 100? **2300**

What is 5972 to the nearest 1000?

What is 994 to the nearest 10?

What is 2349 to the nearest 100?

What is 5511 to the nearest 1000?

What is 2299 to the nearest 100?

What is 179 to the nearest 10?

Year 4 – Number and place value
- *Round any number to the nearest 10, 100 or 1000.*

Rounding to 10, 100 or 1000 (3)

Learning objectives
I can round numbers to the nearest 10, 100 or 1000.

+ 3 ×
¾ = 8
1.5 ÷

To solve the joke, round each number and write the answer in the circle. Then use the grid to find the letter that goes with each answer and write it on the line. The first one is done for you!

5000	3500	2140	6200	800	6670	370	1000
M	S	I	P	C	A	R	E

What is 5228 to the nearest 1000? (5000)

What is 2136 to the nearest 10? ( )

What is 829 to the nearest 100? ( )

What is 1158 to the nearest 1000? ( )

What is 795 to the nearest 10? ( )

What is 366 to the nearest 10? ( )

What is 811 to the nearest 1000? ( )

What is 6674 to the nearest 10? ( )

What is 4912 to the nearest 1000? ( )

Year 4 – Number and place value
- *Round any number to the nearest 10, 100 or 1000.*

Rounding to 10, 100 or 1000 (4)

Learning objectives
I can round numbers to the nearest 10, 100 or 1000.

+ 3 ×
¾ = 8
1.5 ÷

To solve the joke, round each number and write the answer in the oval. Then use the grid to find the letter that goes with each answer and write it on the line. The first one is done for you!

5000	3500	2140	6200	800	6670	370	1000
M	S	I	P	C	A	R	E

What is 4522 to the nearest 1000? (5000)

What is 2144 to the nearest 10?

What is 753 to the nearest 100?

What is 725 to the nearest 1000?

What is 804 to the nearest 10?

What is 372 to the nearest 10?

What is 2139 to the nearest 10?

What is 3546 to the nearest 100?

What is 6153 to the nearest 100?

What is 2141 to the nearest 10?

What is 1099 to the nearest 1000?

What is 3495 to the nearest 100?

Year 4 – Number and place value

- *Round any number to the nearest 10, 100 or 1000.*

Roman numerals (1)

Learning objectives
I can understand Roman numerals for numbers up to 100.

+ 3 ×
¾ = 8
1.5 ÷

To solve the joke, change the Roman numeral into a number and write the answer in the circle. Then use the grid to find the letter that goes with each answer and write it on the line. The first one is done for you!

4	50	55	30	9	100	90	17	64	26	53	70
D	N	E	R	I	P	O	C	K	L	S	U

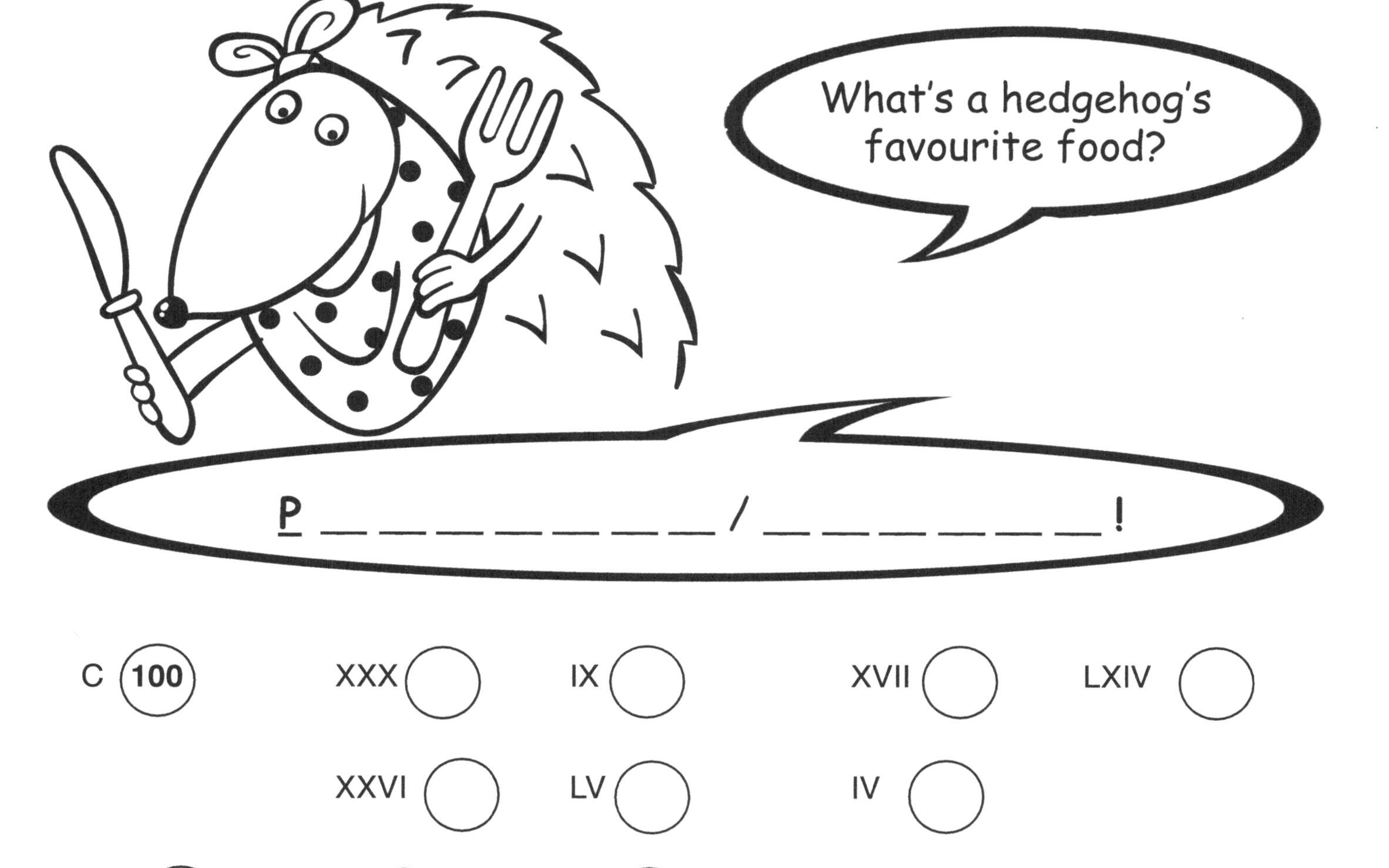

C (100)   XXX ( )   IX ( )   XVII ( )   LXIV ( )

XXVI ( )   LV ( )   IV ( )

XC ( )   L ( )   IX ( )   XC ( )   L ( )

LIII ( )

Year 4 – Number and place value
- *Read Roman numerals to 100 (I to C) and know that over time, the numeral system changed to include the concept of zero and place value.*

Roman numerals (2)

Learning objectives
I can understand Roman numerals for numbers up to 100.

To solve the joke, change the Roman numeral into a number and write the answer in the circle. Then use the grid to find the letter that goes with each answer and write it on the line. The first one is done for you!

4	50	55	30	9	100	90	17	64	26	53	70
D	N	E	R	I	P	O	C	K	L	S	U

Year 4 – Number and place value
- *Read Roman numerals to 100 (I to C) and know that over time, the numeral system changed to include the concept of zero and place value.*

Roman numerals (3)

Learning objectives
I can understand Roman numerals for numbers up to 100.

+ 3 ×
¾ = 8
1.5 ÷

To solve the joke, change the Roman numeral into a number and write the answer in the circle. Then use the grid to find the letter that goes with each answer and write it on the line. The first one is done for you!

97	26	33	51	59	96	40	80	8	19	94	67
G	E	N	A	R	W	C	O	L	Y	T	J

24	69	45	99	54
B	U	I	M	P

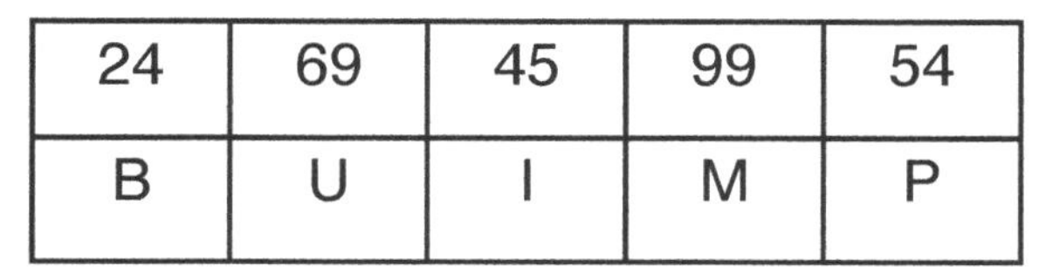

A / _ _ _ _ _ _ _ _ _ / _ _ _ _ _ _ _ _ _ !

LI (51)

XCVI ( ) LXXX ( ) LXXX ( ) VIII ( ) VIII ( ) XIX ( )

LXVII ( ) LXIX ( ) XCIX ( ) LIV ( ) XXVI ( ) LIX ( )

Year 4 – Number and place value

- *Read Roman numerals to 100 (I to C) and know that over time, the numeral system changed to include the concept of zero and place value.*

Roman numerals (4)

Learning objectives
I can understand Roman numerals for numbers up to 100.

+ 3 ×
¾ = 8
1.5 ÷

To solve the joke, change the Roman numeral into a number and write the answer in the circle. Then use the grid to find the letter that goes with each answer and write it on the line. The first one is done for you!

97	26	33	51	59	96	40	80	8	19	94	67
G	E	N	A	R	W	C	O	L	Y	T	J

24	69	45	99	54
B	U	I	M	P

What do you get if you cross a sheep with a radiator?

C _ _ _ _ _ _ _ / _ _ _ _ _ _ _ _ _ _ !

XL (40) XXVI ( ) XXXIII ( ) XCIV ( ) LIX ( ) LI ( ) VIII ( )

XXIV ( ) VIII ( ) XXVI ( ) LI ( ) XCIV ( ) XLV ( ) XXXIII ( ) XCVII ( )

Year 4 – Number and place value

- *Read Roman numerals to 100 (I to C) and know that over time, the numeral system changed to include the concept of zero and place value.*

Addition and subtraction up to 4-digits (1)

Learning objectives	
I can add numbers with up to 4-digits. I can subtract two numbers with up to 4-digits.	+ 3 × ¾ = 8 1.5 ÷

To solve the joke, use a written method to find the answer and write it in the oval. Then use the grid to find the letter that goes with each answer and write it on the line. The first one is done for you!

2250	3219	3637	1445	2873	7628	9189	1122	7358	9881
S	C	O	A	L	U	T	R	H	P

1302 + 948 = **2250**

6824 – 3187 =

5183 + 2445 =

1885 – 763 =

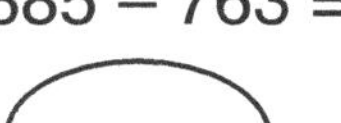

4323 + 5558 =

9255 – 1627 =

466 + 1784 =

6823 – 4573 =

Year 4 – Addition and subtraction

- *Add and subtract numbers with up to 4 digits using the formal written methods of columnar addition and subtraction where appropriate.*

Addition and subtraction up to 4-digits (2)

Learning objectives
I can add numbers with up to 4-digits.
I can subtract two numbers with up to 4-digits.

To solve the joke, use a written method to find the answer and write it in the oval. Then use the grid to find the letter that goes with each answer and write it on the line. The first one is done for you!

2250	3219	3637	1445	2873	7628	9189	1122	7358	9881
S	C	O	A	L	U	T	R	H	P

2209 + 134 + 876 =

3219

7333 – 5888 =

8212 + 977 =

5231 – 3786 =

8113 – 6991 =

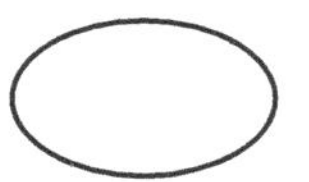

3302 – 2180 =

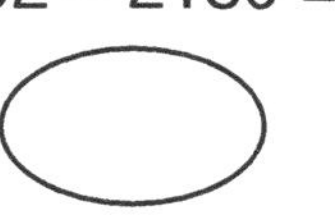

4377 + 2981=

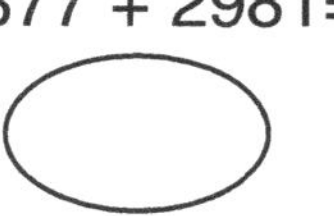

Year 4 – Addition and subtraction

- *Add and subtract numbers with up to 4 digits using the formal written methods of columnar addition and subtraction where appropriate.*

Addition and subtraction up to 4-digits (3)

Learning objectives
I can add numbers with up to 4-digits.
I can subtract two numbers with up to 4-digits.

+ 3 ×
¾ = 8
1.5 ÷

To solve the joke, use a written method to find the answer and write it in the oval. Then use the grid to find the letter that goes with each answer and write it on the line. The first one is done for you!

2644	7212	1856	2689	877	5373	1765	3111	7253	5296
O	K	T	G	H	I	R	E	N	Y

856	3154	3822
A	B	L

5311 – 3455 =

4253 - 3376 =

7105 – 3994 =

1357 + 3939 =

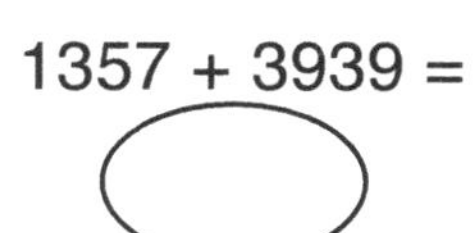

1356 + 873 + 1593 =

1356 + 1288 =

5613 – 2924 =

9124 – 3751 =

4651 + 1877 + 725 =

Year 4 – Addition and subtraction
- *Add and subtract numbers with up to 4 digits using the formal written methods of columnar addition and subtraction where appropriate.*

Addition and subtraction up to 4-digits (4)

Learning objectives
I can add numbers with up to 4-digits.
I can subtract two numbers with up to 4-digits.

To solve the joke, use a written method to find the answer and write it in the oval. Then use the grid to find the letter that goes with each answer and write it on the line. The first one is done for you!

2644	7212	1856	2689	877	5373	1765	3111	7253	5296
O	K	T	G	H	I	R	E	N	Y

856	3154	3822
A	B	L

1754 + 1178 + 222 = **3154**

9554 – 8698 = (   )

5233 – 3468 = (   )

2856 + 4356 = (   )

Year 4 – Addition and subtraction
- *Add and subtract numbers with up to 4 digits using the formal written methods of columnar addition and subtraction where appropriate.*

Tables up to 12 x 12 (1)

Learning objectives
I know or can work out facts for all the times tables up to 12 x 12.
I know or can work out the related divisions for all the times tables up to 12 x 12.

+ 3 ×
¾ = 8
1.5 ÷

To solve the joke, write the answer to the multiplication tables question in the oval. Then use the grid to find the letter that goes with each answer and write it on the line. The first one is done for you!

6	36	144	7	11	56	8	72	96
O	U	L	C	H	I	K	A	G

$49 \div 7 =$

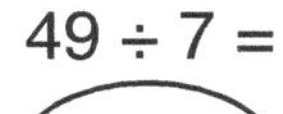
7

$55 \div 5 =$

$7 \times 8 =$

$42 \div 6 =$

$72 \div 9 =$

$8 \times 9 =$

$12 \times 8 =$

$36 \div 6 =$

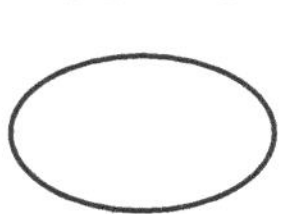

Year 4 – Multiplication and division
- *Recall multiplication and division facts for multiplication tables up to 12 × 12.*

Tables up to 12 x 12 (2)

Learning objectives
I know or can work out facts for all the times tables up to 12 x 12.
I know or can work out the related divisions for all the times tables up to 12 x 12.

+ 3 × ¾ = 8 1.5 ÷

To solve the joke, write the answer to the multiplication tables question in the oval. Then use the grid to find the letter that goes with each answer and write it on the line. The first one is done for you!

6	36	144	7	11	56	8	72	96
O	U	L	C	H	I	K	A	G

6 x 12 = 72

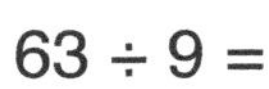

63 ÷ 9 =   9 x 4 =   21 ÷ 3 =   32 ÷ 4 =   42 ÷ 7 =   30 ÷ 5 =

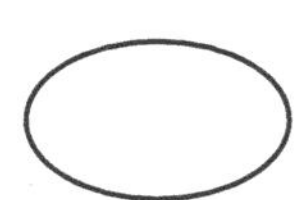
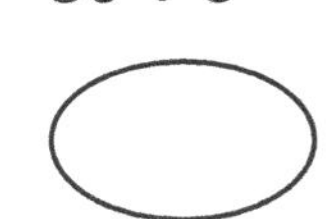

28 ÷ 4 =   12 x 12 =   6 x 6 =   35 ÷ 5 =   48 ÷ 6 =

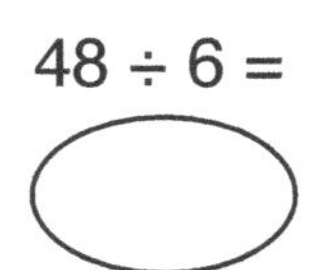

Year 4 –Multiplication and division
- *Recall multiplication and division facts for multiplication tables up to 12 × 12.*

Tables up to 12 x 12 (3)

Learning objectives
I know or can work out facts for all the times tables up to 12 x 12.
I know or can work out the related divisions for all the times tables up to 12 x 12.

To solve the jokes, write the answer to the multiplication tables question in the oval. Then use the grid to find the letter that goes with each answer and write it on the line. The first one is done for you!

5	24	121	4	12	64	81	9	108	35	3
S	H	P	E	I	F	M	A	T	C	L

What do you call a haddock that won't share?

S _ _ _ _ _ - _ _ _ _ _ !

10 ÷ 2 = 5

6 x 4 =

16 ÷ 4 =

15 ÷ 5 =

27 ÷ 9 =

8 x 8 =

72 ÷ 6 =

40 ÷ 8 =

3 x 8 =

What do you call a fish on a table?

_ / _ _ _ _ _ _ _ _ / _ _ _ _ !

81 ÷ 9 =

11 x 11=

24 ÷ 8 =

36 ÷ 4 =

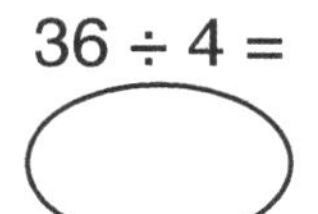

84 ÷ 7 =

5 x 7 =

32 ÷ 8 =

9 x 9 =

54 ÷ 6 =

9 x 12 =

Year 4 – Multiplication and division
- *Recall multiplication and division facts for multiplication tables up to 12 × 12.*

Applying tables knowledge (1)

Learning objectives
I can use place value and tables to multiply multiples of 10 and 100.
I can multiply numbers by 0 and 1.
I can divide numbers by 1.
I can multiply three 1-digit numbers together.

To solve the joke, write the answer to the multiplication tables question in the oval. Then use the grid to find the letter that goes with each answer and write it on the line. The first one is done for you!

12	480	120	30	50	1500	720	0	5600	36	280	11
D	N	H	O	Y	R	T	S	I	E	G	C

What did the Vikings use to communicate?

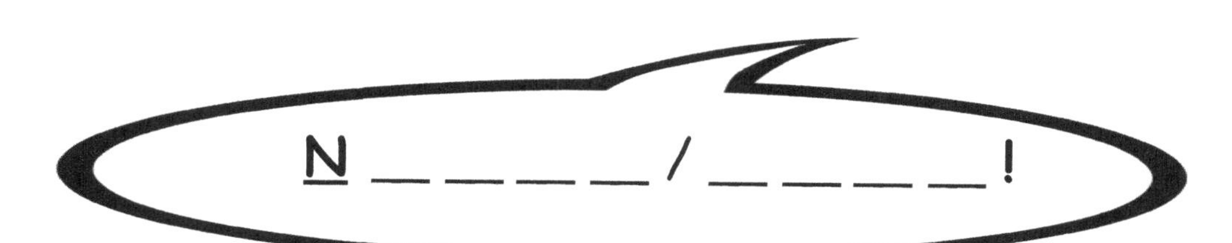

80 x 6 = **480**

3 x 5 x 2 =

3 x 500 =

8 x 0 =

6 x 2 x 3 =

11 x 1 =

6 x 5 x 1 =

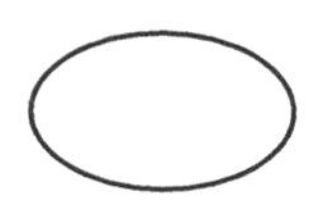

12 ÷ 1 =

1 x 9 x 4 =

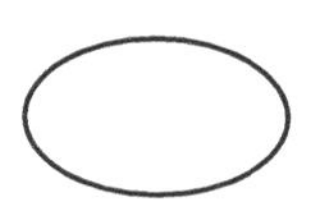

Year 4 – Multiplication and division
- *Use place value, known and derived facts to multiply and divide mentally, including:*
  - *multiplying by 0 and 1*
  - *dividing by 1*
  - *multiplying together three numbers.*

Applying tables knowledge (2)

Learning objectives
I can use place value and tables to multiply multiples of 10 and 100.
I can multiply numbers by 0 and 1.
I can divide numbers by 1.
I can multiply three 1-digit numbers together.

To solve the joke, write the answer to the multiplication tables question in the oval. Then use the grid to find the letter that goes with each answer and write it on the line. The first one is done for you!

12	480	120	30	50	1500	720	0	5600	36	280	11
D	N	H	O	Y	R	T	S	I	E	G	C

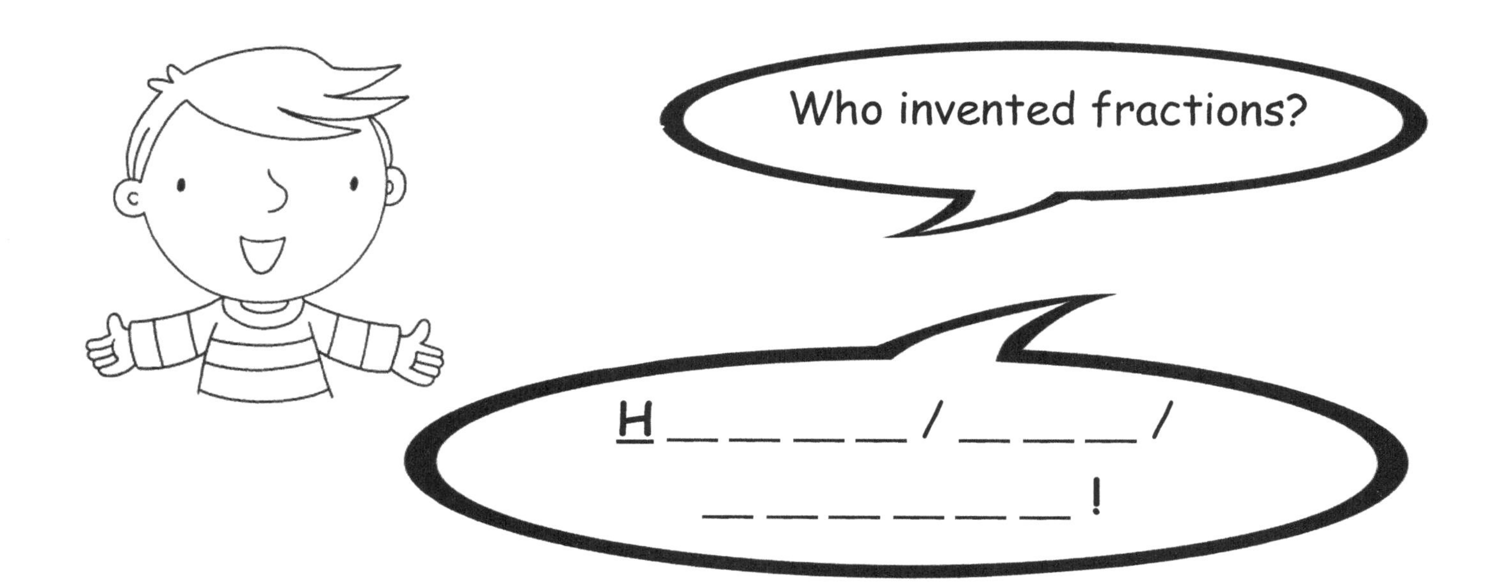

10 x 6 x 2 =

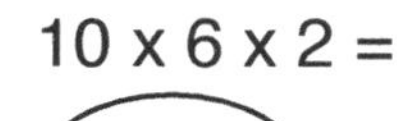

3 x 3 x 4 =

60 x 8 =

300 x 5 =

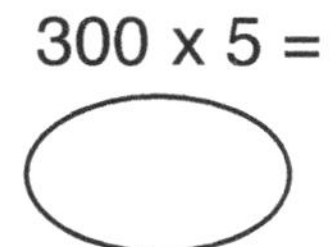

5 x 5 x 2 =

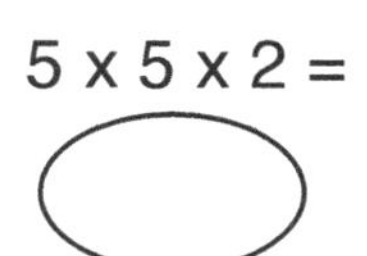

80 x 9 =

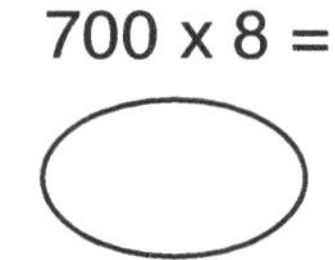

4 x 3 x 10 =

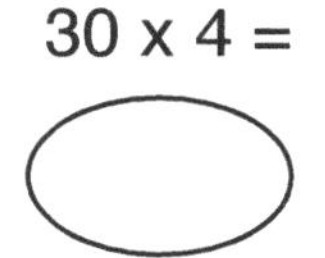

3 x 1 x 12 =

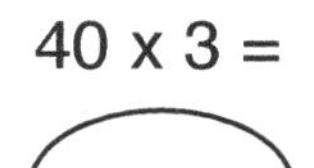

9 x 2 x 2 = 700 x 8 = 70 x 4 = 30 x 4 = 60 x 12 = 40 x 3 =

Year 4 – Multiplication and division

- *Use place value, known and derived facts to multiply and divide mentally, including:*
  - *multiplying by 0 and 1*
  - *dividing by 1*
  - *multiplying together three numbers.*

Applying tables knowledge (3)

Learning objectives

I can use place value and tables to multiply multiples of 10 and 100.
I can multiply numbers by 0 and 1.
I can divide numbers by 1.
I can multiply three 1-digit numbers together.

+ 3 ×
¾ = 8
1.5 ÷

To solve the joke, write the answer to the multiplication tables question in the oval. Then use the grid to find the letter that goes with each answer and write it on the line. The first one is done for you!

200	360	180	40	80	0	450	2000	480	24	45	4000	9	8
O	A	T	P	S	E	W	R	M	I	B	N	H	Z

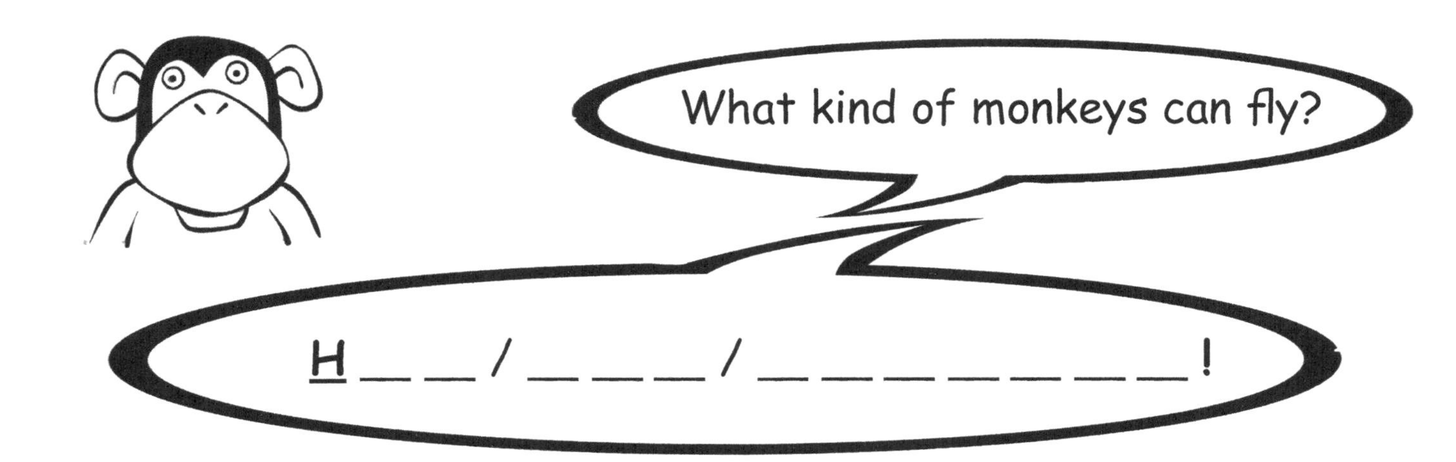

9 x 1 =

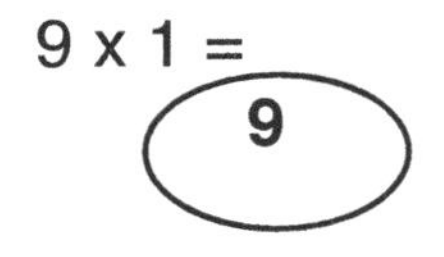

40 x 5 =

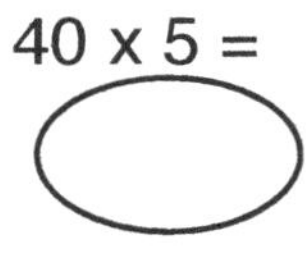

60 x 3 =

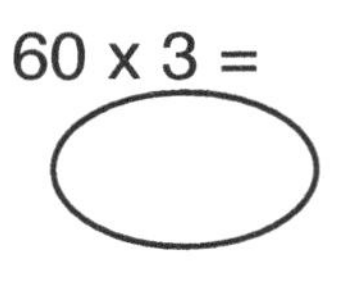

90 x 4 =

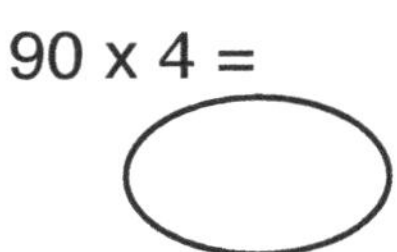

2 x 4 x 3 =

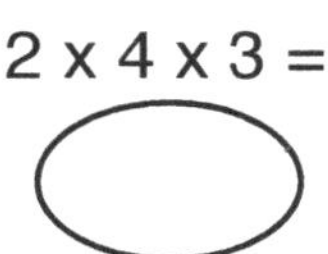

500 x 4 =

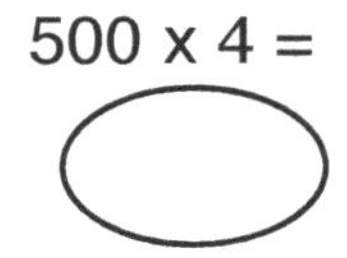

3 x 3 x 5 =

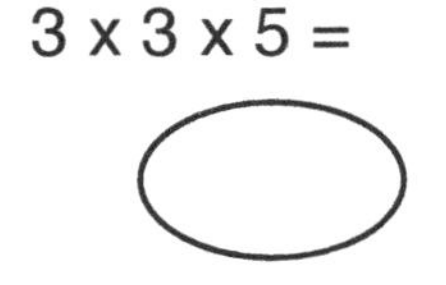

6 x 60 =

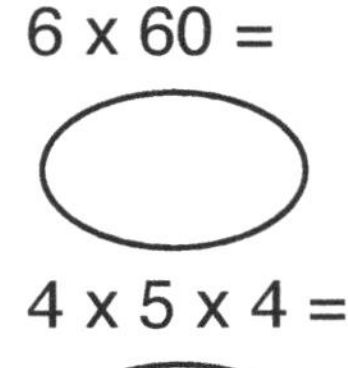

3 x 5 x 3 =

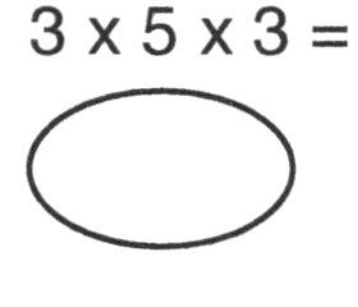

50 x 4 =

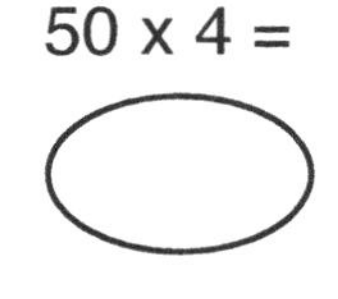

4 x 5 x 10 =

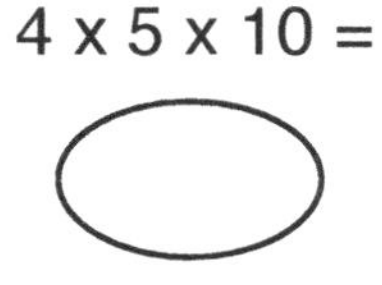

500 x 8 =

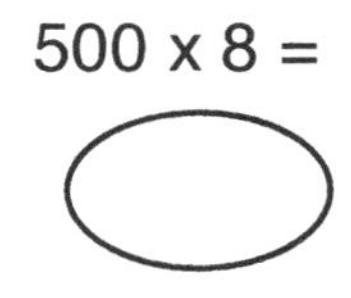

4 x 5 x 4 =

Year 4 – Multiplication and division

- *Use place value, known and derived facts to multiply and divide mentally, including:*
  - *multiplying by 0 and 1*
  - *dividing by 1*
  - *multiplying together three numbers.*

Applying tables knowledge (4)

Learning objectives
I can use place value and tables to multiply multiples of 10 and 100.
I can multiply numbers by 0 and 1.
I can divide numbers by 1.
I can multiply three 1-digit numbers together.

To solve the joke, write the answer to the multiplication tables question in the oval. Then use the grid to find the letter that goes with each answer and write it on the line. The first one is done for you!

200	360	180	40	80	0	450	2000	480	24	45	4000	9	8
O	A	T	P	S	E	W	R	M	I	B	N	H	Z

What do you call a monkey who's good at swimming?

A / _ _ _ _ _ - _ _ _ _ _ _ _ _ !

30 x 12 = **360**

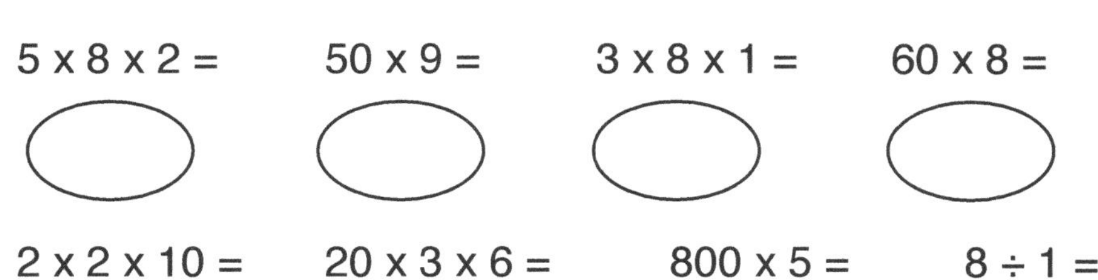

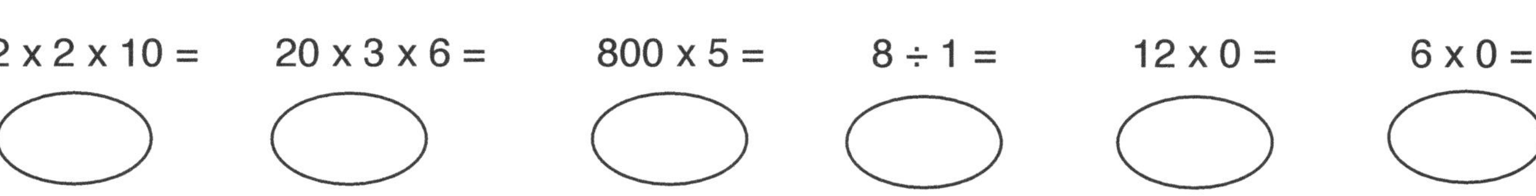

Year 4 – Multiplication and division
- *Use place value, known and derived facts to multiply and divide mentally, including: multiplying by 0 and 1; dividing by 1; multiplying together three numbers.*

Multiplying by 2 and 3 digit numbers (1)

Learning objectives
I can multiply a 2-digit number by a 1-digit number.
I can multiply a 3-digit number by a 1-digit number.

To solve the joke, use a written method to work out the answer to the question. Write the answer in the oval, then use the grid to find the letter that goes with each answer and write it on the line. The first one is done for you!

232	834	285	2412	240	2640	1410	1272	184	4508	156	308
N	H	T	R	E	S	L	U	P	M	A	B

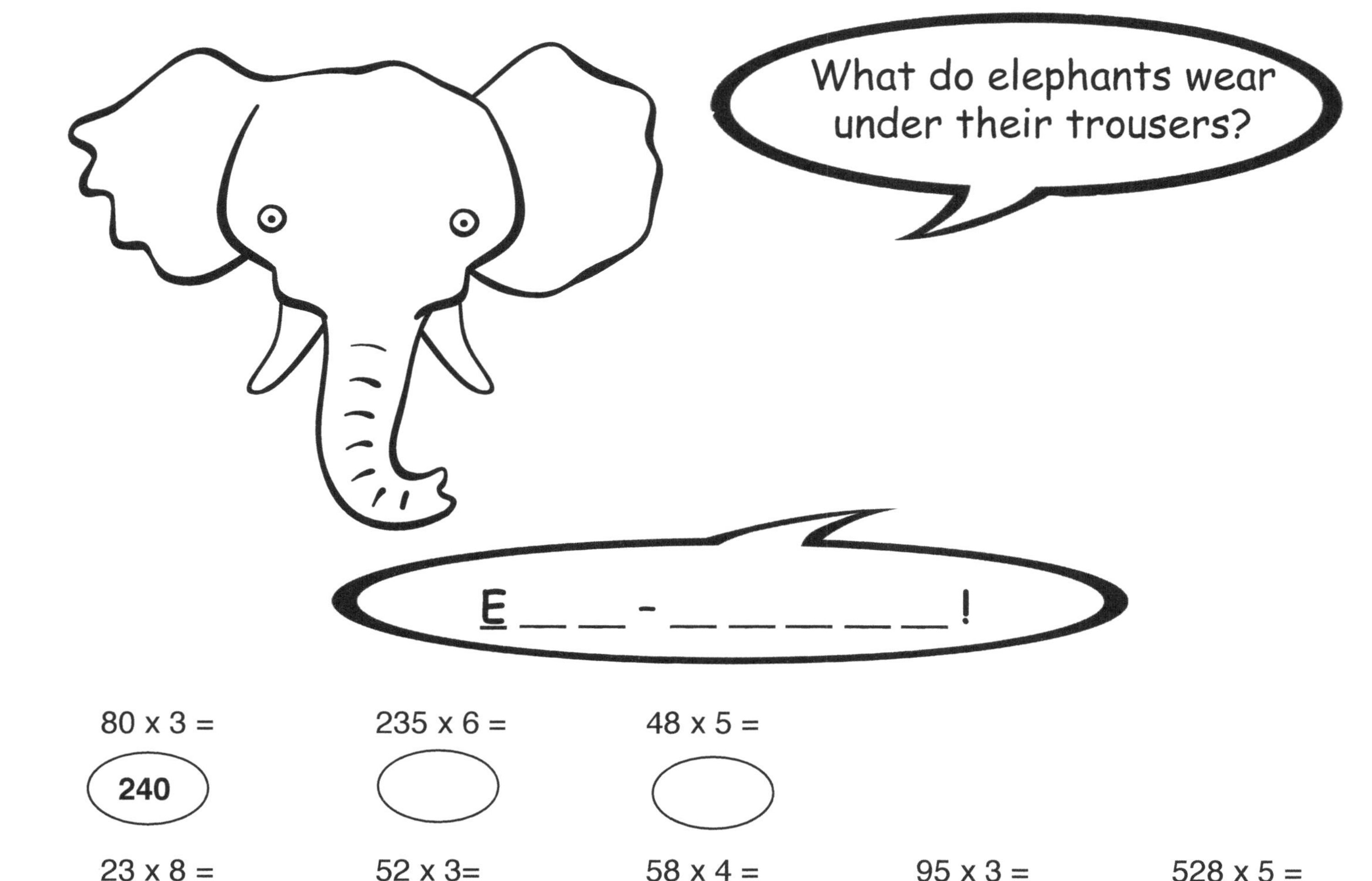

80 x 3 = **240**

235 x 6 =

48 x 5 =

23 x 8 =

52 x 3=

58 x 4 =

95 x 3 =

528 x 5 =

Year 4 – Multiplication and division
- *Multiply two-digit and three-digit numbers by a one-digit number using formal written layout.*

Multiplying by 2 and 3 digit numbers (2)

Learning objectives
I can multiply a 2-digit number by a 1-digit number.
I can multiply a 3-digit number by a 1-digit number.

To solve the joke, use a written method to work out the answer to the question. Write the answer in the oval, then use the grid to find the letter that goes with each answer and write it on the line.

232	834	285	2412	240	2640	1410	1272	184	4508	156	308
N	H	T	R	E	S	L	U	P	M	A	B

What are big, grey and protect you from the rain?

U _ - _ _ _ _ _ _ - _ _ _ _ _ _ !

318 x 4 = 1272 644 x 7 =

77 x 4 = 804 x 3 = 30 x 8 = 282 x 5 = 470 x 3 = 39 x 4 =

92 x 2 = 139 x 6 = 78 x 2 = 29 x 8 = 57 x 5 = 660 x 4 =

 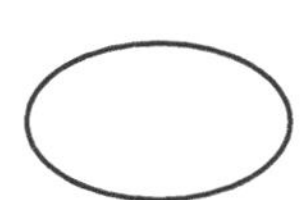 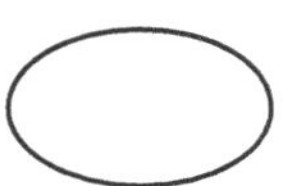 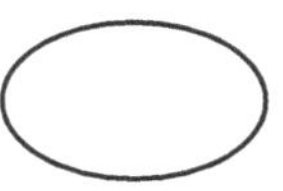  

Year 4 – Multiplication and division
- *Multiply two-digit and three-digit numbers by a one-digit number using formal written layout.*

Multiplying by 2 and 3 digit numbers (3)

Learning objectives
I can multiply a 2-digit number by a 1-digit number.
I can multiply a 3-digit number by a 1-digit number.

To solve the joke, write the answer to the multiplication tables question in the oval. Then use the grid to find the letter that goes with each answer and write it on the line. The first one is done for you!

623	368	3948	464	495	1602	5238	534	4923	2520	468
W	G	T	N	L	O	F	M	I	E	S

46 x 8 =
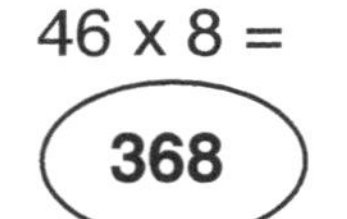

58 x 8 =

267 x 6 =

89 x 6 =
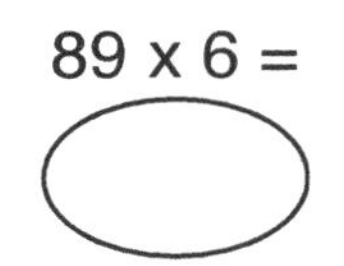

630 x 4 =
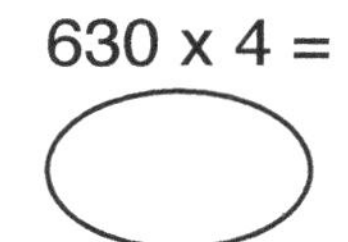

78 x 6 =

89 x 7 =

315 x 8 =

420 x 6 =

564 x 7 =

92 x 4 =

116 x 4 =

534 x 3 =

178 x 3 =

504 x 5 =

Year 4 – Multiplication and division
- *Multiply two-digit and three-digit numbers by a one-digit number using formal written layout.*

Multiplying by 2 and 3 digit numbers (4)

Learning objectives
I can multiply a 2-digit number by a 1-digit number.
I can multiply a 3-digit number by a 1-digit number.

To solve the joke, write the answer to the multiplication tables question in the oval. Then use the grid to find the letter that goes with each answer and write it on the line. The first one is done for you!

623	368	3948	464	495	1602	5238	534	4923	2520	468
W	G	T	N	L	O	F	M	I	E	S

840 x 3 = 2520

55 x 9 =

873 x 6 =

547 x 9 =

156 x 3 =

Year 4 – Multiplication and division
- *Multiply two-digit and three-digit numbers by a one-digit number using formal written layout.*

Equivalent fractions (1)

Learning objectives
I know some equivalent fractions.
I can use my tables knowledge to help me find equivalent fractions.

To solve the jokes, work out which fraction is the odd one out. Two of them are equivalent to each other and one isn't. Write the odd one out in the circle, then use the grid to find the letter that goes with each answer and write it on the line. The first one is done for you!

$2/5$	$5/8$	$7/9$	$4/9$	$6/8$	$5/6$	$5/5$	$3/7$	$6/9$
C	B	O	A	F	T	I	M	N

What did the vampire do when he caught a cold?

C _ _ _ _ _ _ _ !

$4/5$ $2/5$ $8/10$ (**$2/5$**) $1/3$ $3/9$ $7/9$ ( ) $6/8$ $2/4$ $1/2$ ( )

$3/6$ $4/8$ $6/8$ ( ) $2/3$ $5/5$ $4/6$ ( ) $6/9$ $5/5$ $3/3$ ( )

Who is a vampire's favourite superhero?

$4/5$ $5/8$ $8/10$ ( ) $4/9$ $4/8$ $5/10$ ( ) $3/4$ $5/6$ $6/8$ ( )

$3/7$ $3/9$ $1/3$ ( ) $6/9$ $4/9$ $4/6$ ( ) $6/9$ $1/3$ $2/6$ ( )

Year 4 – Fractions (including decimals)
- *Recognise and show, using diagrams, families of common equivalent fractions.*

Equivalent fractions (2)

Learning objectives
I know some equivalent fractions.
I can use my tables knowledge to help me find equivalent fractions.

To solve the jokes, work out which fraction is the odd one out. Two of them are equivalent to each other and one isn't. Write the odd one out in the circle, then use the grid to find the letter that goes with each answer and write it on the line. The first one is done for you!

4/5	6/8	1/2	8/10	7/8	3/6	1/3	3/7	3/4
E	I	S	M	T	P	C	A	W

What did the cat say when he ran out of money?

I' __ / __ __ __ __ !

3/4 6/8 8/10 ◯

2/6 1/3 3/6 ◯

3/4 3/7 6/8 ◯

3/4 3/6 4/8 ◯

What do cats like to listen to?

__ __ __ __ __ __ __ __ !

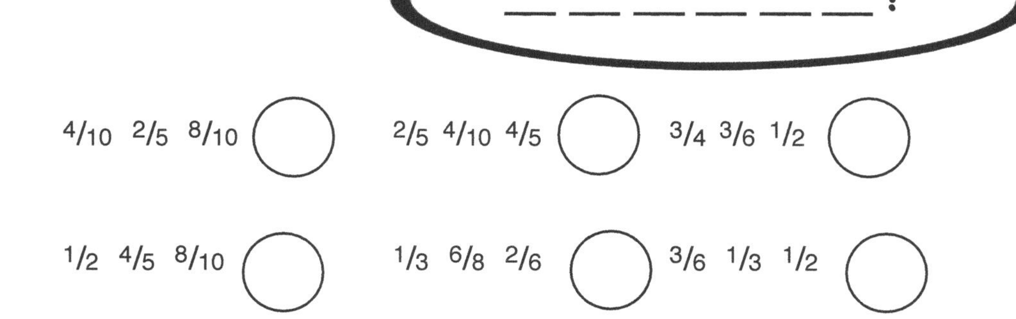

Year 4 – Fractions (including decimals)
- *Recognise and show, using diagrams, families of common equivalent fractions.*

Fractions of numbers (1)

Learning objectives
I can calculate fractions of a quantity.

To solve the joke, answer the maths question and write the answer in the circle. Then, use the grid to find the letter that goes with each answer and write it on the line. The first one is done for you!

16	15	20	32	6	56	12	8	14	10	21
A	G	S	U	P	N	E	Y	L	I	N

$^5/_6$ of 24 = (20) $^2/_3$ of 9 = ( ) $^3/_4$ of 16 = ( ) $^2/_3$ of 21 = ( ) $^1/_2$ of 28 = ( )

$^1/_{10}$ of 100 = ( ) $^3/_4$ of 28 = ( ) $^5/_{100}$ of 300 = ( )

Year 4 – Fractions (including decimals)
- *Solve problems involving increasingly harder fractions to calculate quantities, and fractions to divide quantities, including non-unit fractions where the answer is a whole number.*

Fractions of numbers (2)

Learning objectives
I can calculate fractions of a quantity.

To solve the joke, answer the maths question and write the answer in the circle. Then, use the grid to find the letter that goes with each answer and write it on the line. The first one is done for you!

16	15	20	32	6	56	12	8	14	10	21
A	G	S	U	P	N	E	Y	L	I	N

$2/4$ of 32 = 16

$5/10$ of 40 = ◯

$4/5$ of 40 = ◯

$7/8$ of 64 = ◯

$8/9$ of 63 = ◯

$2/8$ of 32 = ◯

$4/7$ of 35 = ◯

$2/100$ of 300 = ◯

$1/12$ of 144 = ◯

$2/8$ of 56 = ◯

$2/7$ of 49 = ◯

Year 4 – Fractions (including decimals)

- *Solve problems involving increasingly harder fractions to calculate quantities, and fractions to divide quantities, including non-unit fractions where the answer is a whole number.*

Fractions of numbers (3)

Learning objectives
I can calculate fractions of a quantity.

To solve the joke, answer the maths question and write the answer in the circle. Then, use the grid to find the letter that goes with each answer and write it on the line. The first one is done for you!

18	5	30	25	6	36	24	8	7	10	11
N	H	K	E	Y	F	R	A	V	W	O

$\frac{1}{100}$ of 800 = 8

$\frac{2}{10}$ of 50 =

$\frac{1}{6}$ of 66 =

$\frac{3}{7}$ of 42 =

$\frac{1}{10}$ of 300 =

$\frac{3}{100}$ of 200 =

Year 4 – Fractions (including decimals)

- *Solve problems involving increasingly harder fractions to calculate quantities, and fractions to divide quantities, including non-unit fractions where the answer is a whole number.*

Fractions of numbers (4)

Learning objectives I can calculate fractions of a quantity.	+ 3 × ¾ = 8 1.5 ÷

To solve the joke, answer the maths question and write the answer in the circle. Then, use the grid to find the letter that goes with each answer and write it on the line. The first one is done for you!

18	5	30	25	6	36	24	8	7	10	11
N	H	K	E	Y	F	R	A	V	W	O

$\frac{1}{12}$ of 60 = (5) $\frac{2}{4}$ of 16 = ( ) $\frac{1}{100}$ of 600 = ( )

$\frac{4}{7}$ of 63 = ( ) $\frac{1}{4}$ of 100 = ( ) $\frac{1}{3}$ of 21 = ( ) $\frac{5}{8}$ of 40 = ( ) $\frac{6}{100}$ of 400 = ( )

Year 4 – Fractions (including decimals)

- *Solve problems involving increasingly harder fractions to calculate quantities, and fractions to divide quantities, including non-unit fractions where the answer is a whole number.*

Addition and subtraction of fractions (1)

Learning objectives
I can add and subtract fractions where the answer could be more than one whole.

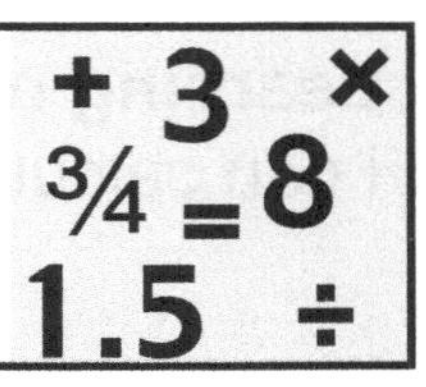

To solve the jokes, work out the answer to the fractions question. Write the answer in the circle, then use the grid to find the letter that goes with each answer and write it on the line. The first one is done for you!

7/10	13/5	11/4	4/9	8/8	8/6	12/5	3/7	9/8
U	A	C	Y	H	O	L	K	B

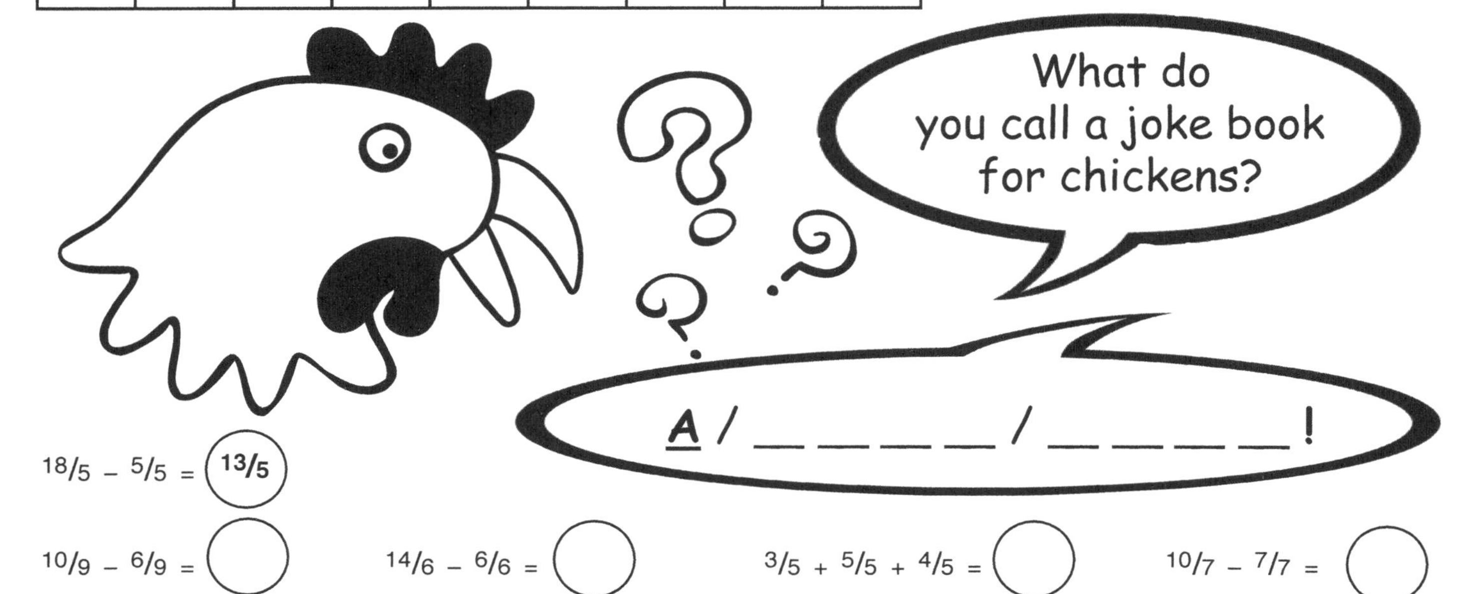

$18/5 - 5/5 =$ (13/5)

$10/9 - 6/9 =$ ◯ $14/6 - 6/6 =$ ◯ $3/5 + 5/5 + 4/5 =$ ◯ $10/7 - 7/7 =$ ◯

$17/8 - 8/8 =$ ◯ $15/6 - 7/6 =$ ◯ $1/6 + 3/6 + 4/6 =$ ◯ $15/7 - 12/7 =$ ◯

What did the chicken say when it laid a square egg?

$20/6 - 12/6 =$ ◯ $15/10 - 8/10 =$ ◯ $3/4 + 6/4 + 2/4 =$ ◯ $19/8 - 11/8 =$ ◯

Year 4 – Fractions (including decimals)
- *Add and subtract fractions with the same denominator*

Addition and subtraction of fractions (2)

Learning objectives
I can add and subtract fractions where the answer could be more than one whole.

+ 3 ×
¾ = 8
1.5 ÷

To solve the jokes, work out the answer to the fractions question. Write the answer in the circle, then use the grid to find the letter that goes with each answer and write it on the line. The first one is done for you!

2/5	11/8	7/9	4/9	8/8	15/6	9/7	3/5	6/9	2/7	9/8
G	I	S	C	P	E	A	B	W	U	R

What button did the bear press to stop the film?

$4/8 + 2/8 + 2/8 =$ (8/8) $2/7 + 5/7 + 2/7 =$ ◯ $12/9 - 6/9 =$ ◯ $12/9 - 5/9 =$ ◯

What do polar bears eat for lunch?

$7/8 + 4/8 =$ ◯ $13/9 - 9/9 =$ ◯ $3/6 + 5/6 + 7/6 =$ ◯

$8/5 - 5/5 =$ ◯ $12/7 - 10/7 =$ ◯ $3/8 + 4/8 + 2/8 =$ ◯ $11/5 - 9/5 =$ ◯

$21/6 - 6/6 =$ ◯ $4/8 + 5/8 =$ ◯ $19/9 - 12/9 =$ ◯

Year 4 – Fractions (including decimals)
- *Add and subtract fractions with the same denominator*

Equivalent fractions and decimals (1)

Learning objectives
I know the equivalent decimals for tenths and hundreths.
I know the decimal equivalents for $1/2$, $1/4$ and $3/4$.

+ 3 ×
¾ = 8
1.5 ÷

To solve the jokes, convert the fractions and decimals and write the equivalent answer in the oval. Then use the grid to find the letter that goes with each answer and write it on the line. The first one is done for you!

$1/2$	0.05	0.5	$6/10$	$1/4$	0.25	$1/10$	0.3	$3/4$	$15/100$
O	P	S	T	I	E	A	M	H	Y

What starts with t, ends with t and is full of t?

T _ _ / _ _ _ _ !

0.6 as a fraction ($6/10$)

$1/4$ as a decimal

0.1 as a fraction

$5/100$ as a decimal

0.5 as a fraction

0.6 as a fraction

What should you do if a bull charges you?

_ _ _ _ / _ _ _ _ !

$5/100$ as a decimal

0.1 as a fraction

0.15 as a fraction

0.75 as a fraction

0.25 as a fraction

$3/10$ as a decimal

Year 4 – Fractions (including decimals)
- *Recognise and write decimal equivalents of any number of tenths or hundredths*
- *Recognise and write decimal equivalents to $1/2$, $1/4$ and $3/4$.*

Equivalent fractions and decimals (2)

Learning objectives
I know the equivalent decimals for tenths and hundreths.
I know the decimal equivalents for $1/2$, $1/4$ and $3/4$.

To solve the jokes, convert the fractions and decimals and write the equivalent answer in the oval. Then use the grid to find the letter that goes with each answer and write it on the line. The first one is done for you!

0.5	$1/2$	0.25	$1/10$	$6/100$	$1/4$	0.75	$6/10$	$7/100$	0.1
Y	A	D	T	F	R	S	P	M	H

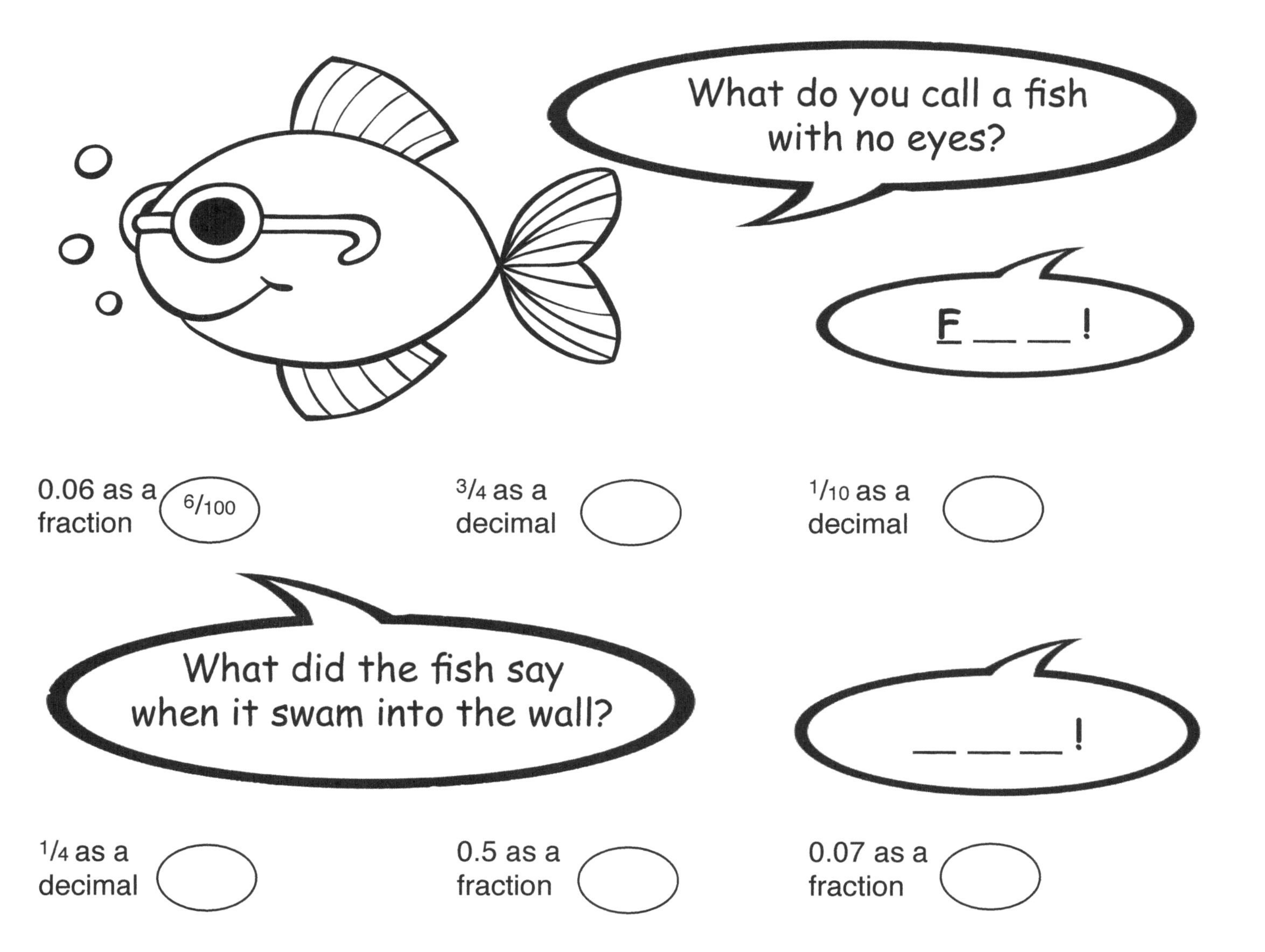

Year 4 – Fractions (including decimals)
- *Recognise and write decimal equivalents of any number of tenths or hundredths*
- *Recognise and write decimal equivalents to $1/2$, $1/4$ and $3/4$.*

Equivalent fractions and decimals (3)

Learning objectives
I know the equivalent decimals for tenths and hundreths.
I know the decimal equivalents for 1/2, 1/4 and 3/4.

To solve the joke, convert the fractions and decimals and write the equivalent answer in the oval. Then use the grid to find the letter that goes with each answer and write it on the line. The first one is done for you!

0.5	1/2	0.05	1/10	6/100	1/4	0.75	6/10	7/100	0.1
Y	A	D	T	F	R	S	P	M	H

0.06 as a fraction (6/100)

0.25 as a fraction ( )

50/100 as a decimal ( )

5/100 as a decimal ( )

0.5 as a fraction ( )

1/2 as a decimal ( )

Year 4 – Fractions (including decimals)

- *Recognise and write decimal equivalents of any number of tenths or hundredths*
- *Recognise and write decimal equivalents to 1/2, 1/4 and 3/4.*

Divide by 10 or 100 (1)

Learning objectives
I can use place value to divide a 1-digit or a 2-digit number by 10 or 100.

To solve the joke, use place value to multiply or divide the numbers by 10 or 100 and write the answer in the oval. Then use the grid to find the letter that goes with each answer and write it on the line. The first one is done for you!

0.63	0.29	2.9	0.84	0.4	8.4	9.2	0.05	0.5	0.7	0.92	0.6	0.07
G	K	M	H	O	L	I	Y	T	N	P	E	S

4 ÷ 10 = 0.4

92 ÷ 10 =

7 ÷ 10 =

29 ÷ 100 =

29 ÷ 10 =

6 ÷ 10 =

70 ÷ 100 =

5 ÷ 10 =

Year 4 – Multiplication and division

- *Find the effect of dividing a one- or two-digit number by 10 and 100, identifying the value of the digits in the answer as ones, tenths and hundredths.*

Divide by 10 or 100 (2)

Learning objectives
I can use place value to divide a 1-digit or a 2-digit number by 10 or 100.

To solve the joke, use place value to multiply or divide the numbers by 10 or 100 and write the answer in the oval. Then use the grid to find the letter that goes with each answer and write it on the line. The first one is done for you!

0.63	0.29	2.9	0.84	0.4	8.4	9.2	0.05	0.5	0.7	0.92	0.6	0.07
G	K	M	H	O	L	I	Y	T	N	P	E	S

50 ÷ 100 =

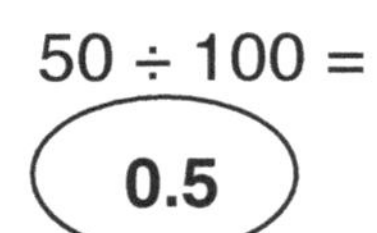

84 ÷ 100 =

60 ÷ 100 =

40 ÷ 100 =

84 ÷ 10 =

5 ÷ 100 =

29 ÷ 10 =

92 ÷ 100 =

92 ÷ 10 =

63 ÷ 100 =

7 ÷ 100 =

Year 4 – Multiplication and division

- *Find the effect of dividing a one- or two-digit number by 10 and 100, identifying the value of the digits in the answer as ones, tenths and hundredths.*

Divide by 10 or 100 (3)

Learning objectives
I can use place value to divide a 1-digit or a 2-digit number by 10 or 100.

To solve the jokes, use place value to multiply or divide the numbers by 10 or 100 and write the answer in the oval. Then use the grid to find the letter that goes with each answer and write it on the line. The first one is done for you!

0.6	5.4	9.3	0.5	0.07	0.9	4.5	3.9	0.7	0.06
S	U	M	O	L	R	Y	A	E	G

What do you get if you hang from a tree in the jungle?

S _ _ _ / _ _ _ _ _ !

6 ÷ 10 = **0.6**   50 ÷ 100 = ____   9 ÷ 10 = ____   70 ÷ 100 = ____

39 ÷ 10 = ____   90 ÷ 100 = ____   93 ÷ 10 = ____   60 ÷ 100 = ____

45 ÷ 10 = ____   5 ÷ 10 = ____   54 ÷ 10 = ____   90 ÷ 100 = ____

7 ÷ 100 = ____   7 ÷ 10 = ____   6 ÷ 100 = ____   60 ÷ 100 = ____

Year 4 – Multiplication and division

- *Find the effect of dividing a one- or two-digit number by 10 and 100, identifying the value of the digits in the answer as ones, tenths and hundredths.*

Rounding to the nearest whole number (1)

Learning objectives
I can round numbers with one decimal place to the nearest whole number.

To solve the joke, round each number to the nearest whole number and write the answer in the circle. Then use the grid to find the letter that goes with each answer and write it on the line. The first one is done for you!

1	2	3	4	5	6	7	8	9	10
L	D	T	P	C	A	O	R	W	Y

2.5 (**3**) 7.1 ( )

3.8 ( ) 7.4 ( ) 1.4 ( ) 0.6 ( ) 9.5 ( )

8.8 ( ) 7.1 ( ) 6.5 ( ) 2.1 ( )

Year 4 – Fractions (including decimals)
- *Round decimals with one decimal place to the nearest whole number.*

Rounding to the nearest whole number (2)

Learning objectives
I can round numbers with one decimal place to the nearest whole number.

To solve the joke, round each number to the nearest whole number and write the answer in the circle. Then use the grid to find the letter that goes with each answer and write it on the line. The first one is done for you!

1	2	3	4	5	6	7	8	9	10
L	D	T	P	C	A	O	R	W	Y

6.3 (**6**)

4.5 ( ) 5.8 ( ) 8.2 ( ) 7.6 ( ) 6.8 ( ) 2.7 ( )

Year 4 – Fractions (including decimals)
- *Round decimals with one decimal place to the nearest whole number.*

Rounding to the nearest whole number (3)

Learning objectives
I can round numbers with one decimal place to the nearest whole number.

To solve the joke, round each number to the nearest whole number and write the answer in the circle. Then use the grid to find the letter that goes with each answer and write it on the line. The first one is done for you!

1	8	2	9	3	10	4	11	5	12	6	7
T	O	R	E	N	G	I	A	C	W	K	L

1.1 (1) 1.8 ( ) 4.3 ( ) 4.9 ( ) 6.2 ( )

8.4 ( ) 2.1 ( )

0.5 ( ) 11.6 ( ) 8.8 ( ) 8.5 ( ) 0.7 ( )

Year 4 – Fractions (including decimals)
- *Round decimals with one decimal place to the nearest whole number.*

Rounding to the nearest whole number (4)

Learning objectives
I can round numbers with one decimal place to the nearest whole number.

+ 3 ×
¾ = 8
1.5 ÷

To solve the joke, round each number to the nearest whole number and write the answer in the circle. Then use the grid to find the letter that goes with each answer and write it on the line. The first one is done for you!

1	8	2	9	3	10	4	11	5	12	6	7
T	O	R	E	N	G	I	A	C	W	K	L

10.8 (11)   2.9 ( )

3.9 ( )   6.5 ( )   7.4 ( )

8.7 ( )   11.4 ( )   9.8 ( )   6.7 ( )   8.6 ( )

Year 4 – Fractions (including decimals)
- *Round decimals with one decimal place to the nearest whole number.*

Comparing and ordering decimals (1)

Learning objectives
I can show that I understand the place value of decimals by finding the largest one.

To solve the joke, work out which is the largest decimal number in each group and write it in the oval. Then use the grid to find the letter that goes with each answer and write it on the line. The first one is done for you!

2.12	2.19	1.8	2.9	1.7	2.99	2.91	2.1	1.6	2.5
W	S	A	P	I	N	T	K	E	R

2.11	1.8	1.7	2.19	1.1	2.5	2.11
2.09	1.7	1.5	2.91	1.3	2.3	2.09
2.07	1.5	1.6	1.99	1.6	2.1	2.12
2.12	1.6	1.4	2.12	1.5	2.2	2.19
2.12						

Year 4 – Fractions (including decimals)
- *Compare numbers with the same number of decimal places up to two decimal places.*

Comparing and ordering decimals (2)

Learning objectives
I can show that I understand the place value of decimals by finding the largest one.

To solve the joke, work out which is the largest decimal number in each group and write it in the oval. Then use the grid to find the letter that goes with each answer and write it on the line. The first one is done for you!

2.12	2.19	1.8	2.9	1.7	2.99	2.91	2.1	1.6	2.5
W	S	A	P	I	N	T	K	E	R

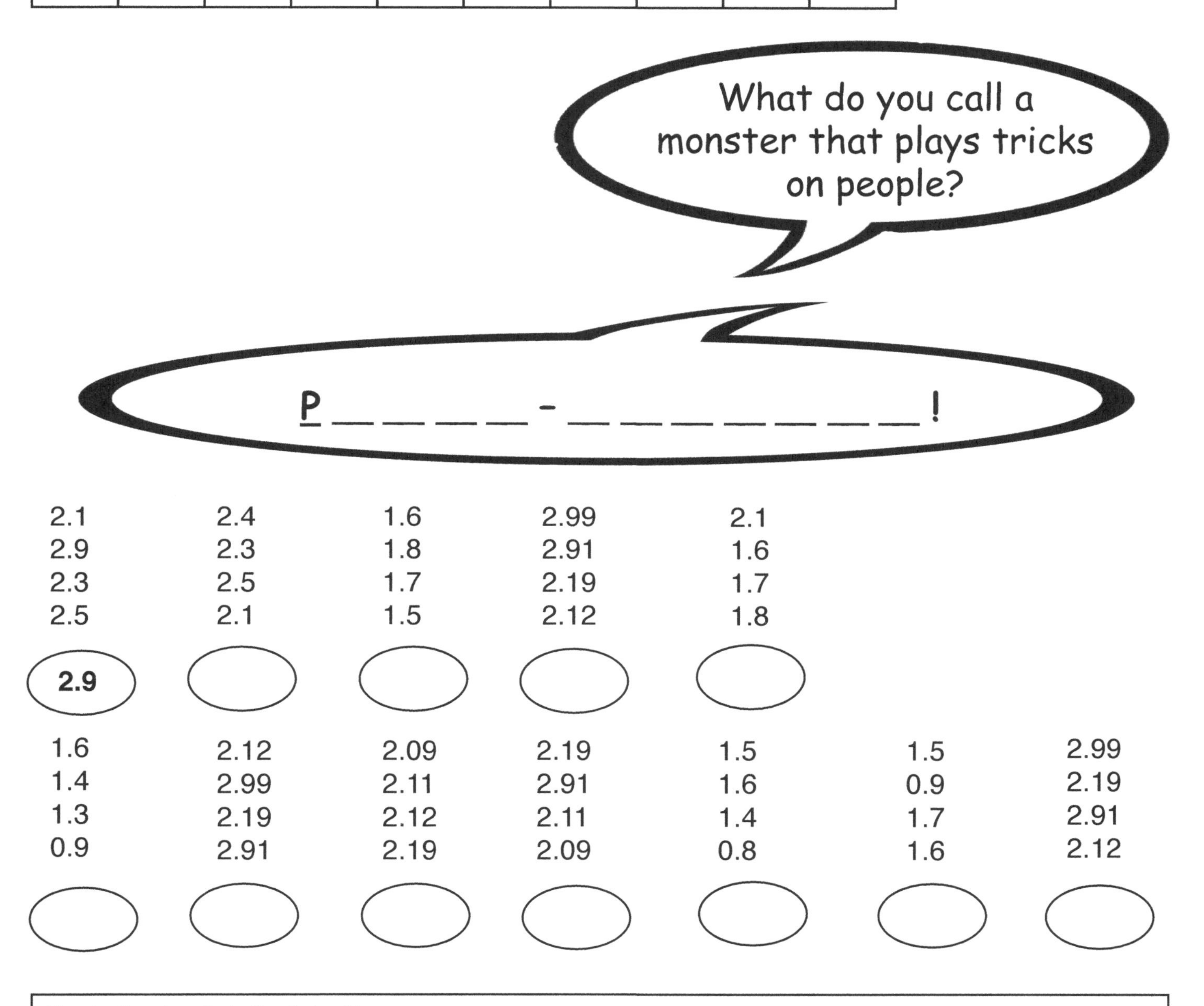

Year 4 – Fractions (including decimals)
- *Compare numbers with the same number of decimal places up to two decimal places*

Comparing and ordering decimals (3)

Learning objectives
I can show that I understand the place value of decimals by finding the largest one.

To solve the joke, work out which is the largest decimal number in each group and write it in the oval. Then use the grid to find the letter that goes with each answer and write it on the line. The first one is done for you!

4.8	5.56	4.6	5.65	5.5	5.6	5.66	5.55	4.5	4.1
R	E	T	A	S	Y	H	U	O	B

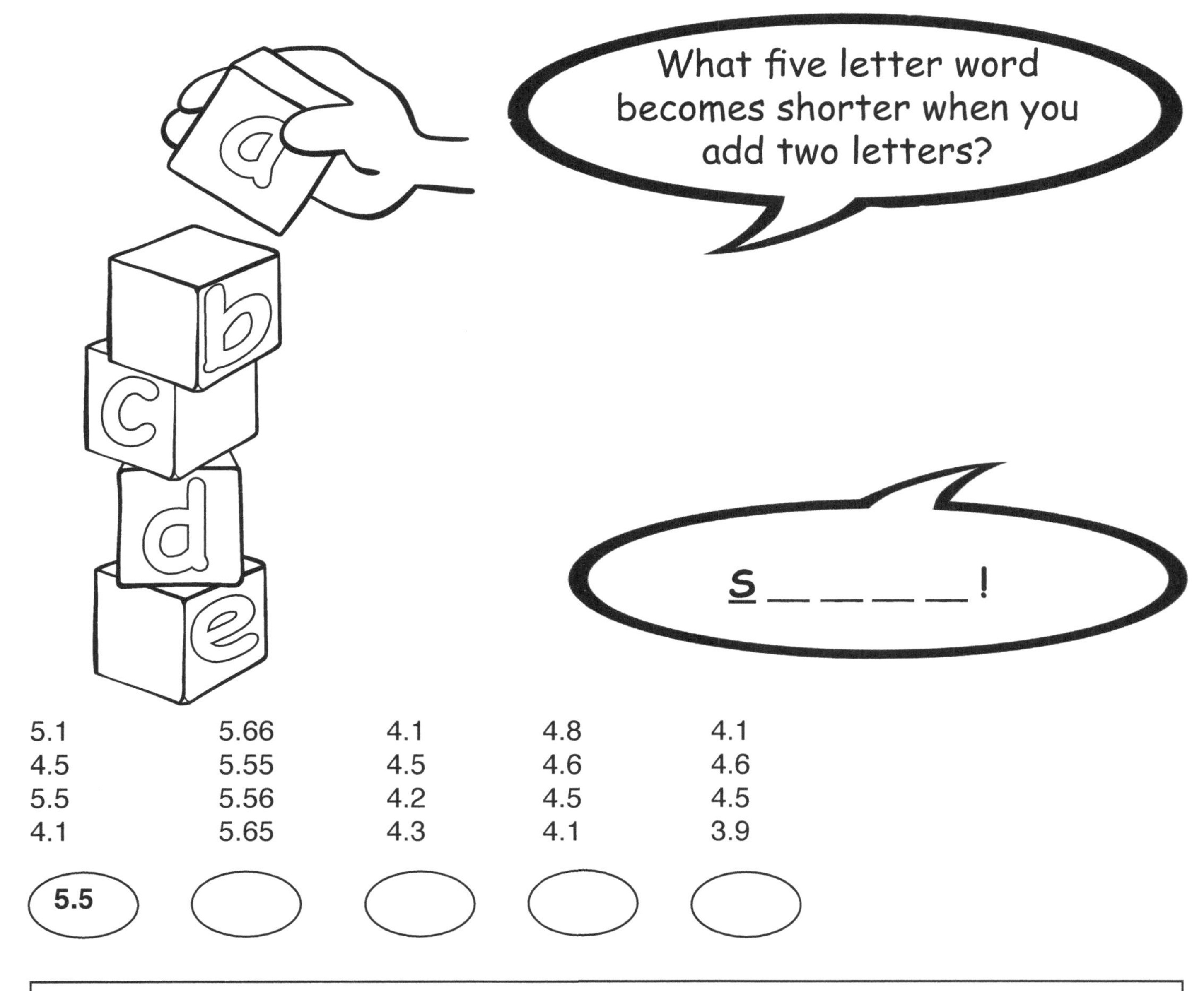

5.1	5.66	4.1	4.8	4.1
4.5	5.55	4.5	4.6	4.6
5.5	5.56	4.2	4.5	4.5
4.1	5.65	4.3	4.1	3.9
5.5				

Year 4 – Fractions (including decimals)

- *Compare numbers with the same number of decimal places up to two decimal places*

Comparing and ordering decimals (4)

Learning objectives
I can show that I understand the place value of decimals by finding the largest one.

To solve the joke, work out which is the largest decimal number in each group and write it in the oval. Then use the grid to find the letter that goes with each answer and write it on the line. The first one is done for you!

4.8	5.56	4.6	5.65	5.5	5.6	5.66	5.55	4.5	4.1
R	E	T	A	S	Y	H	U	O	B

What gets harder to catch when you run faster?

Y _ _ _ _ / _ _ _ _ _ _ _ _ !

5.5
5.6
4.1
4.8

5.6

4.5
4.3
4.1
4.4

5.16
5.05
5.55
5.06

4.6
4.1
4.8
4.5

4.1
3.5
3.8
2.9

4.5
4.8
4.1
4.6

5.56
5.55
4.69
5.51

5.55
5.56
5.65
4.99

4.6
4.1
3.7
4.5

5.66
5.55
5.65
5.56

Year 4 – Fractions (including decimals)
- *Compare numbers with the same number of decimal places up to two decimal places*

Answers

Year 3

Sequences: 4, 8, 50 and 100

Activity	Answers		Page
Activity 1	250, 24, 104, 500, 72, 750, 64, 200, 104, 400, 16 400, 250, 400, 60, 88	WHEN IT'S READ! A WALK!	5
Activity 2	750, 64, 64, 96, 150, 550, 0, 300, 600 550, 0, 300, 850, 24, 32, 300, 48	ALL OF THEM! THEY DREW!	6

Add and subtract 10 or 100

Activity	Answers		Page
Activity 1	258, 819, 82, 782, 595, 395, 458, 782 82, 819, 258, 819, 395, 609	BONE IDLE! NOBODY!	7
Activity 2	491, 534, 252, 607, 405, 674, 208, 607, 391, 416	BY HARE MAIL!	8
Activity 3	252, 607, 405, 674, 734, 52, 691, 218	HARE CUTS!	9

Place value hundreds, tens and ones

Activity	Answers		Page
Activity 1	6, 9, 100, 8, 1, 50, 4, 100, 8, 9, 70, 100 1, 40, 300, 100, 100, 5, 30	FISH AND SHIPS! A MUSSEL!	10
Activity 2	70, 50, 800, 100, 50, 30, 3, 90, 300, 40, 2, 100 5, 9, 5, 100, 3, 90, 5, 1, 40	LOTS OF CRUMBS! EYES-CREAM!	11

Reading numbers

Activity	Answers		Page
Activity 1	560, 330, 420, 560, 516, 280, 313, 1000, 281, 560, 218, 468	AN EARTH QUACK!	12
Activity 2	670, 985, 240, 240, 214, 241, 958, 617, 241, 617	BLOOD TESTS!	13

Writing numbers

Activity	Answers		Page
Activity 1	seven hundred and eight; seven hundred and eighty; six hundred and eighteen; seven hundred and eighteen; three hundred and two; three hundred and twenty-two; three hundred and twelve; three hundred and twelve; three hundred and twenty	QUACK EGGS!	14
Activity 2	four hundred and thirty-one; five hundred and fifty; five hundred and fifty; four hundred and thirty-one; five hundred and fifteen; four hundred and thirty; four hundred and thirteen; five hundred and fifteen; five hundred and fifty-one	TOOTH ACHE!	15

Add and subtract ones, tens and hundreds

Activity	Answers		Page
Activity 1	482, 138, 446, 933, 828, 297, 753, 446, 567	BONE CHINA!	16
Activity 2	297, 753, 517, 677, 59, 446, 446, 158, 482, 138, 446, 933	HIS FUNNY BONE!	17
Activity 3	452, 198, 454, 929, 98, 452, 198, 454, 929, 98	ROUGH ROUGH!	18
Activity 4	713, 537, 713, 627, 562, 98, 67, 198, 929	A WATCH DOG!	19

Add and subtract 3-digit numbers

Activity	Answers		Page
Activity 1	608, 477, 477, 748, 959, 234, 579, 557, 608	EGGS-PLODE!	20
Activity 2	1075, 557, 243, 1343, 801, 748, 1621, 202, 1216, 488	A DRUM STICK!	21
Activity 3	821, 133, 1403, 595, 1275, 395, 821, 133, 1252, 1110, 133	HENRY THE APE!	22
Activity 4	1252, 1110, 133, 253, 1275, 693, 1252, 687, 253, 1275	APESY DAISY!	23

3, 4 and 8 times tables

Activity 1	12, 96, 36, 9, 24, 56, 36, 9	THE RULER	24
Activity 2	72, 12, 12, 48, 5, 12, 36, 15, 7, 24, 1, 1, 8	IT TASTED FUNNY!	25
Activity 3	24, 44, 8, 6, 11, 24, 3, 3, 1, 3	HARD CHEESE!	26
Activity 4	24, 64, 6, 3, 44, 96, 6, 1, 33, 48, 3, 44, 16	HIDE AND SQUEAK!	27

Multiply 2-digit by 1-digit numbers

Activity 1	120, 180, 136, 160, 108, 120, 160, 261, 144, 132, 162	BAKED BEINGS!	28
Activity 2	180, 276, 365, 144, 132, 288, 160, 162, 180, 288, 160	A JUNGLE SALE!	29
Activity 3	168, 156, 128, 128, 188, 395, 156, 156, 168	DOLLY-WOOD!	30
Activity 4	260, 252, 260, 204, 252, 280, 240, 200, 136, 240, 136, 240	A BARBIE-QUEUE!	31

Recognise and find fractions

Activity 1	3/8, 3/5, 4/9, 2/5, 2/3, 5/6, 1/2	APE-RONS!	32
Activity 2	7, 6, 2, 9, 8, 5, 4, 1, 3	APE-RICOTS!	33
Activity 3	5/8, 3/4, 3/4, 5/6, 3/8, 1/3	MOO-SIC!	34
Activity 4	4, 5, 6, 3, 2, 2, 1	THE MOOS!	35

Add and subtract fractions

Activity 1	5/6, 3/7, 6/7, 6/7, 5/6, 2/5 7/9, 6/8, 5/5, 3/7, 5/8, 6/7	LITTLE SQUIRT!	36
Activity 2	4/9, 3/7, 5/6, 5/6, 6/9, 6/7, 7/8, 2/5, 7/9, 6/8, 5/5, 3/7, 2/6	BILLY THE SQUID!	37
Activity 3	2/5, 7/9, 6/8, 5/6, 3/7, 6/7, 2/6, 7/9, 7/9	PLAY SKULL!	38
Activity 4	3/7, 1/4, 7/8, 5/8, 7/9, 4/9, 5/5, 6/7, 6/9, 4/9, 1/8, 7/8, 3/7	SHERLOCK BONES!	39

Compare and order fractions

Activity 1	3/4, 4/5, 4/5, 3/7, 1/5, 1/3, 3/4, 6/7, 1/2	ALLEY CATS!	40
Activity 2	6/7, 8/9, 4/7, 3/7, 3/7, 2/3, 4/5, 1/1, 3/5, 5/8, 1/4, 1/1, 1/3, 3/7	THREE BLIND MICE!	41
Activity 3	3/5, 1/4, 8/9, 1/3, 4/7, 3/4, 4/5, 6/7, 5/8	DIET CROAK!	42
Activity 4	7/9, 4/5, 1/3, 4/7, 3/4, 4/5, 6/7, 5/8, 4/5, 6/7	HOT CROAK-OA!	43

Year 4

Sequence multiples of 6, 7, 9, 25, and 1000

Activity 1	90, 1000, 275, 54, 1000, 5000, 84, 54, 1000 54, 49, 72, 108	IT GETS WET! EDAM!	44
Activity 2	24, 56, 90, 72, 42, 42, 650, 81, 90, 10 000 375, 7000, 24	THE MOO-VIES PAT	45

1000 more or less

Activity 1	5247, 4087, 4906, 5547, 2398, 5247, 4087, 2098, 3606, 5606, 5606	THE MOTH-BALL!	46
Activity 2	5547, 3606, 3547, 2098, 4906, 4906 512, 4856, 3578, 8649	MAY-BEE! WAVE!	47
Activity 3	3698, 5720, 9511, 6747, 247, 8524, 1919, 8649, 9511, 5720, 897, 8649, 4856	LONG TIME NO SEA!	48

Counting through negative numbers

Activity 1	-13, -9, -10, -6, -12, -4, -7, -5, -12, -2, -3	DROP IT A LINE!	49
Activity 2	0, -10, -5, -13, -11, -12, -8, -1	GOLD FISH!	50
Activity 3	-2, -3, 0, -7, -1, -8, -9, 0	GRAVI-TEA!	51
Activity 4	-5, -10, -1, -6, -9, -4, -10, -1, -9, -5	SLIME FLIES!	52

Place value: Th, H, T, O

Activity 1	1, 200, 8, 1, 900, 7000, 60, 60, 4000, 30, 3000	A SHAM-POODLE!	53
Activity 2	1, 9000, 60, 30, 30, 80, 3000, 400, 30, 60, 40, 3000, 700	A COLLIE-FLOWER!	54
Activity 3	700, 5000, 9000, 200, 20, 500, 2000, 900, 5, 5, 5000	SLITHER-POOL!	55
Activity 4	50, 70, 500, 50, 200, 20, 500, 2000, 7, 5, 50	A FEATHER-BOA!	56

Rounding to 10, 100 or 1000

Activity 1	5200, 400, 2300, 400, 5200, 7000, 100, 5200, 8000, 570, 6000	ON AN OCTO-BUS!	57
Activity 2	2300, 6000, 990, 2300, 6000, 2300, 180	A SEA-SAW!	58
Activity 3	5000, 2140, 800, 1000, 800, 370, 1000, 6670, 5000	MICE CREAM!	59
Activity 4	5000, 2140, 800, 1000, 800, 370, 2140, 3500, 6200, 2140, 1000, 3500	MICE CRISPIES!	60

Roman numerals

Activity 1	100, 30, 9, 17, 64, 26, 55, 4, 90, 50, 9, 90, 50, 53	PRICKLED ONIONS!	61
Activity 2	100, 90, 30, 64, 70, 100, 9, 50, 55, 53	PORK-UPINES!	62
Activity 3	51, 96, 80, 80, 8, 8, 19, 67, 69, 99, 54, 26, 59	A WOOLLY JUMPER!	63
Activity 4	40, 26, 33, 94, 59, 51, 8, 24, 8, 26, 51, 94, 45, 33, 97	CENTRAL BLEATING!	64

Addition and subtraction up to 4-digits

Activity 1	2250, 3637, 7628, 1122, 9881, 7628, 2250, 2250	SOUR PUSS!	65
Activity 2	3219, 1445, 9189, 1445, 1122, 1122, 7358	CAT-ARRH!	66
Activity 3	1856, 877, 3111, 5296, 3822, 2644, 2689, 5373, 7253	THEY LOG IN!	67
Activity 4	3154, 856, 1765, 7212	BARK!	68

Tables up to 12 x 12

Activity 1	7, 11, 56, 7, 8, 72, 96, 6	CHICK-AGO!	69
Activity 2	72, 7, 36, 7, 8, 6, 6, 7, 144, 36, 7, 8	A CUCKOO CLUCK!	70
Activity 3	5, 24, 4, 3, 3, 64, 12, 5, 24	SHELL-FISH!	71
	9, 121, 3, 9, 12, 35, 4, 81, 9, 108	A PLAICE-MAT!	

Applying tables knowledge

Activity 1	480, 30, 1500, 0, 36, 11, 30,12, 36	NORSE CODE!	72
Activity 2	120, 36, 480, 1500, 50, 720, 120, 36, 36, 5600, 280, 120, 720, 120	HENRY THE EIGHTH!	73
Activity 3	9, 200, 180, 360, 24, 2000, 45, 360, 45, 200, 200, 4000, 80	HOT AIR BABOONS!	74
Activity 4	360, 80, 450, 24, 480, 40, 360, 4000, 8, 0, 0	A SWIM-PANZEE!	75

Multiplying by 2- and 3-digit numbers

Activity 1	240, 1410, 240, 184, 156, 232, 285, 2640	ELE-PANTS!	76
Activity 2	1272, 4508, 308, 2412, 240, 1410, 1410, 156, 184, 834, 156, 232, 285, 2640	UM-BRELLA-PHANTS!	77
Activity 3	368, 464, 1602, 534, 2520, 468, 623, 2520, 2520,	GNOME, SWEET	

Activity	Answers	Joke	Page
	3948, 368, 464, 1602, 534, 2520	GNOME!	78
Activity 4	2520, 495, 5238, 4923, 468	ELFIS!	79

Equivalent fractions

Activity	Answers	Joke	Page
Activity 1	2/5, 7/9, 6/8, 6/8, 5/5, 6/9	COFFIN!	80
	5/8, 4/9, 5/6, 3/7, 4/9, 6/9	BAT-MAN!	
Activity 2	6/8, 8/10, 3/6, 3/7, 3/4	I'M PAW!	81
	8/10, 4/5, 3/4, 1/2, 6/8, 1/3	MEW-SIC!	

Fractions of numbers

Activity	Answers	Joke	Page
Activity 1	20, 6, 12, 14, 14, 10, 21, 15	SPELL-ING!	82
Activity 2	16, 20, 32, 56, 56, 8, 20, 6, 12, 14, 14	A SUNNY SPELL!	83
Activity 3	8, 10, 11, 18, 30, 6	A WONKY!	84
Activity 4	5, 8, 6, 36, 25, 7, 25, 24	HAY FEVER!	85

Addition and subtraction of fractions

Activity	Answers	Joke	Page
Activity 1	13/5, 4/9, 8/6, 12/5, 3/7, 9/8, 8/6, 8/6, 3/7	A YOLK BOOK!	86
	8/6, 7/10, 11/4, 8/8	OUCH!	
Activity 2	8/8, 9/7, 6/9, 7/9	PAWS!	87
	11/8, 4/9, 15/6, 3/5, 2/7, 9/8, 2/5, 15/6, 9/8, 7/9	ICE BURG-ERS!	

Equivalent fractions and decimals

Activity	Answers	Joke	Page
Activity 1	6/10, 0.25, 1/10, 0.05, 1/2, 6/10	TEA POT!	88
	0.05, 1/10, 15/100, 3/4, 1/4, 0.3	PAY HIM!	
Activity 2	6/100, 0.75, 0.1	FSH!	89
	0.25, 1/2, 7/100	DAM!	
Activity 3	6/100, 1/4, 0.5, 0.05, 1/2, 0.5	FRY-DAY!	90

Divide by 10 or 100

Activity	Answers	Joke	Page
Activity 1	0.4, 9.2, 0.7, 0.29, 2.9, 0.6, 0.7, 0.5	OINK-MENT	91
Activity 2	0.5, 0.84, 0.6, 0.4, 8.4, 0.05, 2.9, 0.92, 9.2, 0.63, 0.07	THE OLYM-PIGS!	92
Activity 3	0.6, 0.5, 0.9, 0.7, 3.9, 0.9, 9.3, 0.6	SORE ARMS!	93
	4.5, 0.5, 5.4, 0.9, 0.07, 0.7, 0.06, 0.6	YOUR LEGS!	

Rounding to the nearest whole number

Activity	Answers	Joke	Page
Activity 1	3, 7, 4, 7, 1, 1, 10, 9, 7, 7, 2	TO POLLY-WOOD!	94
Activity 2	6, 5, 6, 8, 8, 7, 3	A CARROT!	95
Activity 3	1, 2, 4, 5, 6, 8, 2, 1, 12, 9, 9, 1	TRICK OR TWEET!	96
Activity 4	11, 3, 4, 7, 7, 9, 11, 10, 7, 9	AN ILL EAGLE!	97

Comparing and ordering decimals

Activity	Answers	Joke	Page
Activity 1	2.12, 1.8, 1.7, 2.91, 1.6, 2.5, 2.19	WAITERS!	98
Activity 2	2.9, 2.5, 1.8, 2.99, 2.1, 1.6, 2.99, 2.19, 2.91, 1.6, 1.7, 2.99	PRANK-ENSTEIN!	99
Activity 3	5.5, 5.66, 4.5, 4.8, 4.6	SHORT!	100
Activity 4	5.6, 4.5, 5.55, 4.8, 4.1, 4.8, 5.56, 5.65, 4.6, 5.66	YOUR BREATH!	101

Assessment checklist

Put a ✔ in the box when the pupil has successfully completed the activity.

Name ______________________ **Class** ______________

Year 3	1	2	3	4
Sequences: 4, 8, 50 and 100				
Add and subtract 10 or 100				
Place value hundreds, tens and ones				
Reading numbers				
Writing numbers				
Add and subtract ones, tens and hundreds				
Add and subtract 3-digit numbers				
3, 4 and 8 times tables				
Multiply 2-digit by 1-digit numbers				
Recognise and find fractions				
Add and subtract fractions				
Compare and order fractions				

Year 4	1	2	3	4
Sequence multiples of 6, 7, 9, 25 and 1000				
1000 more or less				
Counting through negative numbers				
Place value Th, H, T, O				
Rounding to 10, 100 or 1000				
Roman numerals				
Addition and subtraction up to 4-digits				
Tables up to 12 x 12				
Applying tables knowledge				
Multiply by 2- and 3-digit numbers				
Equivalent fractions				
Fractions of numbers				
Addition and subtraction of fractions				
Equivalent fractions and decimals				
Divide by 10 or 100				
Rounding to the nearest whole number				
Comparing and ordering decimals				

Lightning Source UK Ltd.
Milton Keynes UK
UKOW07f1504230815

257368UK00005B/28/P

9 781783 170845